New Discoveries in the Ripening Processes

Edited by Romina Alina Marc and Crina Carmen Mureșan

Published in London, United Kingdom

New Discoveries in the Ripening Processes
http://dx.doi.org/10.5772/intechopen.111017
Edited by Romina Alina Marc and Crina Carmen Mureşan

Contributors
Stefan Toepfl, Oleksii Parniakov, Sam David Hopper, Alamu Emmanuel Oladeji, Ekpereka Oluchukwu Anajekwu, Wasiu Awoyale, Delphine Amah, Rahman Akinoso, Busie Maziya-Dixon, Zulfiya Mukhidova Shabzalovna, Salakhutdin Zakirov Khashimovich, Martín-Ernesto Tiznado-Hernández, Eduardo-Antonio Trillo-Hernández, Miguel-Angel Hernández-Oñate, Jamal Ayour, Mohamed Benichou, Hassnâa Harrak, Maged E. A. Mohammed, Nashi Alqahtani, Muhammad Munir, Romina Alina Marc, Crina Carmen Mureşan, Anamaria Pop, Georgiana Smaranda Marţiş, Andruţa Elena Mureşan, Alina Narcisa Postolache, Florina Stoica, Ioana Cristina Crivei, Ionuţ-Dumitru Veleşcu, Roxana Nicoleta Raţu

© The Editor(s) and the Author(s) 2024
The rights of the editor(s) and the author(s) have been asserted in accordance with the Copyright, Designs and Patents Act 1988. All rights to the book as a whole are reserved by INTECHOPEN LIMITED. The book as a whole (compilation) cannot be reproduced, distributed or used for commercial or non-commercial purposes without INTECHOPEN LIMITED's written permission. Enquiries concerning the use of the book should be directed to INTECHOPEN LIMITED rights and permissions department (permissions@intechopen.com).
Violations are liable to prosecution under the governing Copyright Law.

Individual chapters of this publication are distributed under the terms of the Creative Commons Attribution 3.0 Unported License which permits commercial use, distribution and reproduction of the individual chapters, provided the original author(s) and source publication are appropriately acknowledged. If so indicated, certain images may not be included under the Creative Commons license. In such cases users will need to obtain permission from the license holder to reproduce the material. More details and guidelines concerning content reuse and adaptation can be found at http://www.intechopen.com/copyright-policy.html.

Notice
Statements and opinions expressed in the chapters are these of the individual contributors and not necessarily those of the editors or publisher. No responsibility is accepted for the accuracy of information contained in the published chapters. The publisher assumes no responsibility for any damage or injury to persons or property arising out of the use of any materials, instructions, methods or ideas contained in the book.

First published in London, United Kingdom, 2024 by IntechOpen
IntechOpen is the global imprint of INTECHOPEN LIMITED, registered in England and Wales, registration number: 11086078, 5 Princes Gate Court, London, SW7 2QJ, United Kingdom
Printed in Croatia

British Library Cataloguing-in-Publication Data
A catalogue record for this book is available from the British Library

Additional hard and PDF copies can be obtained from orders@intechopen.com

New Discoveries in the Ripening Processes
Edited by Romina Alina Marc and Crina Carmen Mureşan
p. cm.

This title is part of the Food Science and Nutrition Book Series, Volume 3
Topic: Food Chemistry
Series Editor: Maria Rosário Bronze
Topic Editor: Thuan-Chew Tan

Print ISBN 978-0-85014-126-9
Online ISBN 978-0-85014-127-6
eBook (PDF) ISBN 978-0-85014-128-3
ISSN 2977-8174

We are IntechOpen,
the world's leading publisher of Open Access books
Built by scientists, for scientists

6,800+
Open access books available

183,000+
International authors and editors

195M+
Downloads

156
Countries delivered to

Our authors are among the
Top 1%
most cited scientists

12.2%
Contributors from top 500 universities

WEB OF SCIENCE™

Selection of our books indexed in the Book Citation Index
in Web of Science™ Core Collection (BKCI)

Interested in publishing with us?
Contact book.department@intechopen.com

Numbers displayed above are based on latest data collected.
For more information visit www.intechopen.com

IntechOpen Book Series
Food Science and Nutrition
Volume 3

Aims and Scope of the Series

The significance of food is undeniable, especially in light of the impending challenge facing humanity: ensuring there will be enough food to meet the basic needs of a population expected to reach approximately 10 billion by 2050. These food-related challenges align with some of the United Nations' sustainable development goals, with a target to achieve them by 2030. One thing is certain: food should be not only nourishing and safe but also tailored to the diverse needs of individuals throughout their lifetimes, all while meeting consumers' sensory expectations. Understanding the diverse chemical composition of food, often referred to as biodiversity, and how these components can contribute to human health by considering factors like bioaccessibility, bioavailability, and bioactivity at the organ level, is crucial for grasping and promoting a healthy diet. Thanks to the continuous evolution of analytical methods and interdisciplinary research, significant strides have been made in the field of food science and nutrition.

Meet the Series Editor

Maria Rosário Bronze has been working in Analytical Chemistry since 1986. Her Ph.D. in 1999 contributed to the study of food products using capillary electrophoresis. The main goal of her research since 1999 has been focused on Analytical Chemistry applied mainly to the analysis of foods and by-products of food industry. She conducted research in collaboration with national and international research groups, at iBET and ITQB Technology Division. From 2017 until 2021 she was head of Food & Health Division at iBET and head of the Food Functionality and Bioactives Laboratory. MR Bronze has been an Associate Professor at the Pharmacy Faculty of Lisbon University and head of the Structural Analysis Laboratory since 2012. As a researcher, MR Bronze is a Senior Scientific Advisor at Food & Health Division at iBET and Head of Food Functionality and Bioactives Laboratory at the same Institute, Collaborator at iMED and Researcher at ITQB NOVA. Her current research is focused on quality and beneficial health effects of food components. Gas and liquid chromatography associated with mass spectrometry are used by MR Bronze in the characterization of samples. Sensory evaluation is also an important area of her research. The main food products studied by her are olive tree products (olive, olive oil, leaves), cereals such as maize, legumes (faba bean, pea, chickpea, lentils) fruits (apple, grapes, opuntia ficus), fruit juices and wine, among others. More recently her interests have also involved biodiversity, bioaccessibility, and bioavailability studies on food products and their components, mainly phytochemicals as phenolic compounds, using different analytical tools such as mass spectrometry. As a senior scientific advisor at Food & Health Division at iBET she is involved in different areas: (i) isolation, characterization and formulation of bioactive and functional compounds or extracts from natural sources and wastes from food and other related industries; (ii) pre-clinical assays to provide support to understand health claims related with the beneficial effects of food nutrients/bioactive components; (iii) establishment of analytical methodologies including mass spectrometry state-of-the-art to fully characterize different matrices, from food products, natural extracts or biological fluids (Food Functionality and Bioactives Laboratory).

Meet the Volume Editors

Romina Alina Marc obtained her Ph.D. in Agronomy University from the University of Agricultural Sciences and Veterinary Medicine (USAVM) of Cluj-Napoca, Romania, in 2015. She is a Doctor of Engineering and university lecturer in the Faculty of Food Science and Technology, USAVM Cluj-Napoca, responsible for research activity in plant food quality control, rheology in the food industry, quality management systems, and food safety. She has published more than 110 journal articles on plant science, innovative food product development, food safety, and traceability of bioactive compounds during processing. She has also published several books and book chapters. She has participated in numerous conferences and research projects. She is the recipient of several national and international awards.

Crina Carmen Mureșan graduated with a degree in chemical engineering from the University Babes Bolyai in Cluj-Napoca, Romania, in 1994, working in a quality control laboratory as a chemist. She. obtained her Ph.D. at the University of Agricultural Sciences and Veterinary Medicine in Cluj-Napoca, Romania. She joined the Faculty of Food Science and Technology in 2003 and is currently teaching food quality control and food safety at the University of Agricultural Sciences and Veterinary Medicine (USAVM) Cluj-Napoca, Romania. She has been a doctoral coordinator in biotechnology since 2021. Dr. Muresan has participated in multiple projects on food engineering, product development, consumer preferences, and food quality control. She and her coauthors have presented more than thirty papers and posters at international conferences and published more than ninety papers in peer-reviewed journals as well as two international book chapters. Her research has on focused the quality assessment (gas chromatography-mass spectrometry, spectrometry) of food. Her published papers address the application of advanced methods for the study of phenolic compounds, volatile compounds, and risk chemicals (organochlorine pesticides) present in food.

Contents

Preface	XIII

Chapter 1   1
An Overview of Ripening Processes
by Romina Alina Marc, Crina Carmen Mureșan, Anamaria Pop, Georgiana Smaranda Marțiș, Andruța Elena Mureșan, Alina Narcisa Postolache, Florina Stoica, Ioana Cristina Crivei, Ionuț-Dumitru Veleșcu and Roxana Nicoleta Rațu

Chapter 2   21
The First Signal to Initiate Fruit Ripening is Generated in the Cuticle: An Hypothesis
by Miguel-Angel Hernández-Oñate, Eduardo-Antonio Trillo-Hernández and Martín-Ernesto Tiznado-Hernández

Chapter 3   33
Impact of Ripening and Processing on Color, Proximate and Mineral Properties of Improved Plantain (*Musa spp AAB*) Cultivars
by Ekpereka Oluchukwu Anajekwu, Alamu Emmanuel Oladeji, Wasiu Awoyale, Delphine Amah, Rahman Akinoso and Busie Maziya-Dixon

Chapter 4   53
Environmentally Friendly Plant Terpenoids and Their Biological Activity
by Salakhutdin Zakirov Khashimovich and Zulfiya Mukhidova Shabzalovna

Chapter 5   71
Cell Wall Enzymatic Activity Control: A Reliable Technique in the Fruit Ripening Process
by Jamal Ayour, Hasnaâ Harrak and Mohamed Benichou

Chapter 6   81
Artificial Ripening Technologies for Dates
by Maged Mohammed, Nashi K. Alqahtani and Muhammad Munir

Chapter 7   107
Impact of PEF (Pulsed Electric Fields) on Olive Oil Yield and Quality
by Oleksii Parniakov, Sam David Hopper and Stefan Toepfl

Preface

Today, consumers demand safe, fresh food in any season and with superior sensory qualities. These requirements involve controlling the ripening processes for fruits, vegetables, processed foods, and enzymes.

Ripening is the process by which foods (fruits, vegetables, processed foods) change their aroma, quality, color, taste, and other textural properties. Ripening is associated with changes in composition, for example, the conversion of starch to sugar or various enzymatic processes.

Following the ripening process, foods become sweeter (e.g., in the case of fruits), tastier, more aromatic, and more nutritious.

To achieve these results in a controlled manner, agronomic engineers, food industry engineers, and scientists use state-of-the-art technologies.

This book, *New Discoveries in the Ripening Processes*, presents a comprehensive overview of the food ripening process.

We would like to thank Ms. Maja Bozicevic and Ms. Mirena Čalmić and the rest of the staff at IntechOpen for their support throughout the publishing process. We also wish to thank the contributing authors who contributed to this book. We would also like to thank our families for their understanding and our colleagues at the University of Agricultural Sciences and Veterinary Medicine Cluj-Napoca, Romania, for their support.

Romina Alina Marc and Crina Carmen Mureșan
Faculty of Food Science and Technology,
Food Engineering Department,
University of Agricultural Sciences and Veterinary Medicine Cluj-Napoca,
Cluj Napoca, Romania

Chapter 1

An Overview of Ripening Processes

Romina Alina Marc, Crina Carmen Mureșan, Anamaria Pop, Georgiana Smaranda Marțiș, Andruța Elena Mureșan, Alina Narcisa Postolache, Florina Stoica, Ioana Cristina Crivei, Ionuț-Dumitru Veleșcu and Roxana Nicoleta Rațu

Abstract

The chapter aims to address an overview of the new discoveries regarding the methods of ripening processes. The chapter presents the latest methods used in fruit and vegetable ripening processes, ripening processes in the food industry, enzymatic ripening processes, and artificial ripening processes. Nowadays everyone wants all kinds of food to be available in every season. Naturally, we find fruits and vegetables in their ripening season, but in order to provide the population with fruit out of season, we import them from different countries, which are not harvested at full maturity, and different adjuvant ripening methods are used. Processed foods are also subjected to ripening processes, the most used being cheese and meat products. These foods are some of the most valued foods nowadays, they are considered luxury products with superior nutritional and taste properties. To achieve these ripening processes, enzymatic processes or artificial ripening are also involved. The purpose of using these processes is to provide consumers with fresh out-of-season food or food with a high degree of sensory and nutritional properties, and at the same time with a superior degree of quality and safety, because the safety of the consumer comes first.

Keywords: ripening fruits, ripening food, enzymatic ripening processes, artificial ripening processes, ripening vegetables

1. Introduction

Fruit ripening is a complex developmental process that includes significant changes in texture, color, flavor, smell, nutrient metabolism, and other quality characteristics that ultimately make the fruit attractive, desirable, and edible to consumers [1, 2]. In fruit ripening, hormonal regulation plays a crucial role, and ethylene is the main hormone involved in the process of triggering and accelerating the ripening process of many fruits [3]. Also, temperature, light, and humidity can influence the fruit ripening process. Thus, higher temperatures can accelerate ripening, while lower temperatures delay the process [4]. Fruit ripening is based on crucial metabolic processes, such as changes in carbohydrate metabolism, hormone synthesis, stress response, or defense mechanisms. All these mechanisms are involved in the stimulation or inhibition of certain genes and proteins [5]. In addition to the previously mentioned, defense mechanisms against pathogens and responses to oxidative stress are also involved in fruit ripening. These processes help to protect the fruits from the action of pathogens and to maintain their quality during the ripening process [6]. Depending on the fruit type, they may have specific characteristics and ways of ripening. In the case of climacteric fruits, such as tomatoes and peaches, they undergo a rapid increase in respiration and ethylene production during ripening, while non-climacteric fruits, such as strawberries, have a gradual ripening process [5, 7].

The production of fruits involves a meticulous selection of practices, ranging from the conventional ones like pruning and thinning, which are essential for supporting healthy growth and optimal fruit maturation, to the application of nowadays irrigation and fertilization techniques that provide appropriate moisture levels and nutrient supply. Furthermore, the integration of pest and disease control methods using sustainable strategies like integrated pest management underscores the industry's priority for environmental protection, also ensuring the food safety of the consumers. For example, pruning is a widespread and essential technique used to enhance the quality of fruits and to maintain the health of plants. Pruning shapes not only the canopy but also enhances light penetration and air circulation within the plant, both of which are crucial for photosynthesis. Furthermore, shaping the plant and improving its phytosanitary health, productivity, and fruit quality makes this activity essential for several crops [8]. Plant thinning at early stages, especially before the cell division is completed, is important for optimizing the density of fruits and producing both good quality fruits and high yields. Since fruits require a substantial utilization of plant carbohydrates and mineral nutritional resources [9] in the first stage of fruit development, this method basically prevents resource competition among fruits. Of all the techniques used in fruit production, hand and chemical thinning are the most prevalent [10].

It is well known that crop growth is heavily dependent on the availability of nutrients and water. Research on precision fertilization and irrigation has long been an important area of study, and with the progress of science and technology, precision fertilization and irrigation have been considerably developed. In this regard, conventional irrigation systems can be substituted with innovative and smart irrigation strategies that can further enhance crop yield. In dry areas, drip irrigation is a suitable alternative due to its low-rate water supply and efficient water savings. On the other hand, in horticulture, sprinkler systems are extensively utilized as they do not necessitate big pipes and disperse water droplets across the area, like rainfall [11]. Fertilizers are necessary for increasing crop productivity, preserving soil fertility, and ensuring a consistent supply of essential nutrients for plant needs. According to Jariwala H. et al., the utilization of controlled-release fertilizers can mitigate nutrient loss, increase nutrient utilization efficiency, and improve soil health [12].

Alongside methods that provide the foundation for robust and optimal fruit development, as well as strategies for irrigation and fertilization, it is crucial to prioritize sustainable approaches to controlling pests and diseases. This is critically important for both environmental protection and ensuring food safety. Integrated pest management includes a wide range of strategies to mitigate the harmful effects caused by pests. Pesticide application is considered as the final resort when alternative methods such as biological, behavioral, or other management strategies have proven ineffective in maintaining pest populations below a specified threshold, leading to substantial crop damage. Integrated pest management of fruit trees is an example of the utilization of several strategies to reduce primary pests while preserving natural enemies that offer indirect or secondary pest control [13].

The emergence and development of genetic engineering, along with precision agriculture, have fundamentally shifted the limits of what may be achieved in horticulture. In addition to improving the productivity and efficacy of fruit cultivation, these technologies also provide novel approaches to address the limitations arising from a continually changing global climate and an increasing population. Digitization in this industry includes various technical innovations, including drone technology, robotics, Internet of Things-based automation, and smartphone applications. These technologies consistently monitor, evaluate, and control soil conditions, water supplies, and weather changes [14]. Preserving the quality of ripened fruits requires proper handling and storage after harvesting. This stage integrates a range of good practices and technologies designed to preserve the fruit's freshness, prevent spoilage, and modulate the fruit's ripening process postharvest. For efficient management, understanding of physiological changes during harvesting is a priority [15]. The fruits continue to breathe, ripen, and go through specific metabolic changes even after they have been harvested. Therefore, extending shelf life and assuring quality requires effective management of these changes. Ideal timing and methods for harvesting are key elements of postharvest handling practices, as they effectively mitigate losses and promote optimal fruit maturation. Additionally, fruits are sorted according to their quality, size, and maturity and cleaned to remove soil and detritus after harvesting [16, 17]. Particularly for specific fruits, rapidly cooling after their harvesting lowers respiration and delays the process of ripening, which is essential for these types of fruits. An extensively used method in this field involves storing the fruit in a controlled atmosphere, where the levels of oxygen, carbon dioxide, and humidity are adjusted. This method aims to prolong the fruit's shelf life and delay the ripening process. Furthermore, since ethylene serves as a catalyst for the ripening process, it is important to remove or regulate its presence in storage facilities for fruits that are susceptible to this hormone. In addition to low temperatures and proper handling, modified-atmosphere packaging is a preservation method that can further minimize the physiological and microbiological degradation of perishable fruits [18, 19]. The postharvest processing and packaging of ripe fruits prolong not only their shelf life but also improve their value by converting them into various forms while maintaining their nutritional and sensory attributes. Fruits are frequently preserved to prolong their shelf life. This method involves cleaning and, in certain situations, heat treatment of the fruits prior to their packing in hermetically sealed containers. The aim of heat treatment is to destroy the microorganisms, although this operation has the drawback of negatively impacting the texture and some heat-sensitive nutrients [18]. By drying/dehydrating fruits by the sun or mechanically, microbial growth and enzymatic reactions are inhibited. Although the objective of this method is to enhance flavor concentration, it comes with the loss of valuable nutrients, particularly vitamins [20]. Freezing produce, on the other hand, effectively maintains the nutritional value and flavor of the fruits. A potential drawback of this approach is the appearance of ice crystals under certain circumstances, which can result in cellular structure impairment and subsequent texture degradation during the thawing process [21]. In the context of packaging, modified atmosphere packaging implies the act of modifying the inside conditions to slow down the rate of ripening and spoilage of fruits [22]. Edible coatings represent another method used to apply a protective barrier to fruits, thereby regulating gas exchange and diminishing water loss. Antimicrobial agents may be specifically incorporated into edible coatings [23]. Increasingly prevalent is smart packaging, involving sensors and indicators that deliver data regarding the ripening and spoilage of fruits [24]. In relation to the impact of processing on the ripening process of fruits, modifications can appear in texture, flavor, nutritional value, or chemical agents. Thus, the flavor and texture of fruits can be influenced by the processing techniques, which determine whether the fruits become softer or harder. Although processing can result in the depletion of specific nutrients, it can also enhance the bioavailability of others, as illustrated by the increased availability of lycopene in tomatoes that have undergone processing [25]. The processing and packing of ripened fruits are crucial for converting them into varied edible forms, prolonging their shelf life, and ensuring their year-round availability. However, the optimization of these processes is difficult as it is necessary to preserve the sensory and nutritional proprieties of the fruit, which are crucial for the well-being and satisfaction of customers.

2. New discoveries in fruits and vegetables ripening processes

The postharvest area of vegetables includes an important part of agricultural production. Fresh produce suffers changes in various characteristics between harvest and consumption, resulting in a significant drop in total quality. The postharvest and storage changes of plants can result in food waste and major economic damage. Approximately 24% of all vegetables produced globally are lost in developed countries during the postharvest process and 50% in developing countries [26]. The degree of postharvest losses depends on environmentally friendly agents, techniques, commodities, as well as the season and location of production. However, investments in high-tech postharvest management practices are still a challenge for low-income countries, where politics with efforts for nonhazardous and safe food management are taking place gradually [27].

Vegetables ripening and senescence are degradative processes focused on metabolic disruption and cellular disintegration mediated by genetically regulated programs [28]. Moreover, the vegetables ripening involves oxidative as well as hydrolytic degradation [29]. Tomato fruit has evolved into a model system for understanding the fundamentals of ripening and quality, due to five features: (1) short life cycle plant, (2) collection of mutants affected in ripening, (3) genomic resources, (4) ripening program regulated by ethylene, and (5) was accompanied by increased H_2O_2 concentrations and the oxidation of lipids and proteins [28, 30].

The control of vegetables ripening is essential to maintain quality and to reduce the losses during the postharvest shelf-life. A crucial factor determining the postharvest qualities of fresh vegetables and the potential to keep these postharvest qualities, is the biology of the parent plant [26]. Respiration rate and cold sensitivity are two physiological characteristics with significant postharvest implications. Plants that have higher rates of respiration tend to have a more limited postharvest life. Respiration is crucial in the postharvest life of fresh produce because it shows the metabolic processes of the tissue [31]. Plants' inherent cold sensitivity is genetically encoded. Cold sensitivity of vegetables constraints our ability to use low-temperature storage [26].

Ethylene relation to postharvest quality of vegetables produced, were subject to several reviews [19, 22, 32–36]. Ethylene (C_2H_4) is a naturally present plant hormone which is linked to postharvest quality maintenance in storage. Postharvest-related processes in which ethylene activity is present are: texture changes, taste, development of physiological disorders, bioactive value, and fruit–pathogen interactions [26]. One of the main challenges in food packaging for modifying ethylene biosynthesis is the use of appropriate regulatory elements (silica gel, clays, zeolite, or activated carbon) for optimal postharvest benefit. The impregnation inside the packaging matrix with ethylene scavenging agents (contain catalysts to enhance in situ oxidation) has caused a reduction in the amount of ethylene scavenging agents. The results of this procedure have been proven to increase the product's shelf life while preserving physical quality and freshness [19].

The occurrence of abscission has a large postharvest implication, as well as during the storage of fresh products. Abscission in vegetables depends on the type of plants and it is necessary to either inhibit or induce [26].

Low-temperature postharvest storage is widely used because it slows down the cell metabolism rate and delays senescence and ripening processes. However, low temperatures may induce a disorder "chilling injury" of vegetables [37]. The cell membrane is regarded to be the area of initial events leading to chilling injuries that occur because of membrane structure and permeability (lipids undergo enzymatic peroxidation).

The use of emerging technology for future improvement of postharvest quality and reduction of losses could potentially have a profound effect on supplying the growing demand for food. Currently, it seems that there are some main approaches: efficiently distribute professional knowledge about postharvest to disadvantaged areas but with high potential plant production potential; The other strategy is to utilize emerging technology for the development of new crop varieties, thus becoming more adapted to postharvest storage [32–34].

3. New discoveries in ripening food processes

Ripened foods derived from animals can be categorized into two primary groups: meat products, which encompass dry-cured pieces and dry-cured fermented goods often prepared through mincing and stuffing techniques, and dairy products, which primarily consist of ripened cheeses. These conventional food items are widely recognized and held in high regard globally [38].

The ripening process of animal-derived products is influenced by environmental circumstances that promote the growth of various microbial communities, which significantly contribute to their transformation. The majority of these microorganisms, including some molds, yeasts, gram-positive catalase-positive cocci (GCC+), and lactic acid bacteria (LAB), exert a beneficial influence on the formation of the desired sensory attributes. State that the appearance of these beneficial bacteria is not exclusive to these goods, as they are usually exposed to the indigenous microbiota of the processing environment [38–40].

Nowadays, various efficacious treatments can be employed in the preservation of matured animal-derived food products. These treatments encompass physical techniques like heat treatments, ionizing radiation, and high hydrostatic pressures, as well as chemical preservatives such as organic acids, antifungal compounds, nitrates, and nitrites. Nevertheless, it should be noted that these techniques may not always be congruent with the natural ripening process and may potentially compromise the sensory attributes of the end product. This is mostly due to their non-selective nature, which can result in the detriment of the beneficial bacteria present in these matured food items [41]. Furthermore, there have been reports indicating that the improper or excessive utilization of various synthetic chemicals for the purpose of managing pathogenic bacteria may contribute to the development of resistance among these organisms. In addition, there is a current consumer preference for clean-label items that are devoid of chemical additives and preservatives [42]. Consequently, the implementation of preventative measures in the management of pathogenic or spoilage microorganisms in ripened foods of animal origin relies on the utilization of biocontrol agents (BCAs) derived from either microbial sources or plant-based sources, such as essential oils (EOs) and spices. It is imperative that these treatments have minimal environmental consequences and demonstrate a neutral or favorable effect on the sensory attributes of matured animal-derived food products. This is particularly crucial, as the organoleptic qualities of such products are clearly defined and greatly valued by customers [40].

3.1 Cheese ripening

Cheese is a category of dairy products that undergo fermentation, utilizing milk as the primary ingredient. It encompasses a diverse range of cheese varieties, characterized by their distinct flavors and physical compositions, which are found across the globe. Each geographical region contributes to the development of its own unique cheese products, influenced by cultural practices and available resources. Cheese can be considered a biocomplex ecosystem that is inhabited by a vast array of microorganisms, referred to as cheese flora. These microorganisms are introduced into the cheese through raw milk, as well as starter and adjunct cultures. The flora present in cheese has a significant role in determining the sensory characteristics of various cheese kinds. This is due to their intricate interaction with milk proteins, carbohydrates, and lipids, which mostly takes place during the crucial cheese manufacturing process known as "ripening" [43].

Therefore, the assessment of cheese quality can be conducted by the examination of ripening levels and the corresponding flavor components [43]. Technological incentives exist to optimize the rate of cheese ripening and save costs, potentially impacting the taste characteristics and overall quality of cheese. The variability in cheese qualities may also be attributed to the uncontrolled proliferation and interaction of cheese flora [44].

To make cheese, ripening is an important technological step that involves a series of biochemical and microbiological events regulated by the metabolic flow of main and adjunct cultures [45]. This process needs to be looked into in more detail so that cheese products can be made with better, more consistent quality at the lowest cost and with the most customer acceptance. Changes that happen during ripening decide the organoleptic quality of cheese. The level of aging is very important for shaping the smell and taste of cheese because it changes the chemicals that make it up [46, 47].

However, the process of ripening significantly influences the evolution of cheese flavor, as it encompasses a sequence of modifications in the composition of cheese, including the formation of fatty acids and the metabolism of lactose. The extent of maturation significantly influences the formation of sensory characteristics in cheese, as a result of many biochemical processes such as proteolysis, glycolysis, and lipolysis that take place during this period.

During the initial phase of ripening, the proteolytic enzyme chymosin acts with αs1-casein, leading to its destruction. The extent of this degradation plays a crucial role in determining the textural characteristics of the cheese. The utilization of robust promoters and protein engineering in biotechnological experiments is being explored to enhance the expression of chymosin. These efforts aim to optimize the production of cheese and maximize its flavor and texture attributes [48].

Numerous elements have been identified as having an influence on the quality of cheese. When these factors are carefully handled and monitored, they contribute to the development of desirable flavors. Conversely, if they are not properly addressed, they can lead to the presence of undesirable off-flavors. Numerous research have been undertaken to investigate the factors that can potentially modify the chemical and physical composition of cheese, owing to the delicate balance between flavor production and the occurrence of off-flavors. The examination of flavor, aroma, taste, and chemical constituents has been employed in the process of recognizing and distinguishing various varieties of cheese. Although it is possible that expert panelists may not always get to a definitive agreement, alternative methods are utilized to generate more tangible outcomes. In this particular scenario, the significance of expert judgments in identifying quality qualities of cheese is acknowledged, alongside the conduction of acceptability assessments by customers [46, 47].

The evaluation and surveillance of cheese maturation pose significant difficulties, yet are of utmost importance due to the complex nature of cheese as a multifactorial biological system. This system comprises many classes of substances (such as lipids, proteins, and carbohydrates) inside a complex physical matrix. The utilization of complementing sensory and analytical methodologies is highly sought after in order to adequately investigate the numerous biochemical alterations that occur during this process. In addition, several technologies including infrared (IR) technology, electronic nose (E-nose), and optical techniques such as computer vision and digital image analysis continue to hold significant potential for the analysis and quality control of cheese, as well as the estimation of its shelf life [47].

3.2 Meat and meat product ripening

The process of ripening meat/aging meat encompasses a series of modifications that occur within the muscle tissue of an animal following its slaughter. These processes lead to alterations in the meat's color, softness, and aroma. The metabolic processes that transpire during the age of meat primarily arise from endogenous enzymes, resulting in glycolysis, proteolysis, and lipolysis. During the process of glycolysis, glucose undergoes metabolic reactions resulting in the production of lactic acid. This metabolic pathway leads to a decrease in muscle pH and the depletion of ATP, which serves as the energy reserves. Energy deprivation results in the breakdown of myofibrillar proteins through the enzymatic activity of endopeptidases and exopeptidases. The proteolysis of meat is significantly influenced by endogenous proteases, namely calpains, cathepsins, and calpastatin. However, it is important to note that exogenous proteases, such as peptidylpeptidases, aminopeptidases, and carboxypeptidases, which are secreted by microorganisms participating in meat fermentation, also play a role in elevating the levels of peptides and amino acids [49]. Another chemical process that occurs with the aging of beef is known as lipolysis, which involves the breakdown of fats in both the muscle and adipose tissue [50].

4. New discoveries in enzymatic ripening processes

Fruits and vegetables are an important element of the human diet because they provide vitamins, minerals, antioxidants, carbohydrates, and fiber. The nutritional quality of fruits is heavily dependent on the ripening stage, and optimal ripeness is recommended for ingestion [51]. Moreover, fruit ripening, qualitative qualities, and numerous physiological changes are all heavily influenced by enzymatic activities. Fruits contain enzymes, which function as biological catalysts that accelerate biochemical operations and are essential to many important fruit functions.

Understanding the basic mechanisms governing fruit and vegetable growing, maturation and ripening is necessary to manipulate fruit and vegetable yield and quality [52, 53].

The following are some of the well-known enzymatic activities in fruits and modified components:

4.1 Cell wall degradation

A key component of the cell wall, pectin is broken down by pectinase enzymes, which also include pectin methylesterase and polygalacturonase. During ripening, this process makes the fruit softer. Moreover, fruit softening is facilitated by the enzymes cellulase and hemicellulose, which break down the cellulose and hemicellulose in the cell wall [54].

Fruit softening is primarily brought on by alterations in the content and structure of the cell walls, which affect the fruit's flavor, texture, scent, and appearance [55]. Xyloglucan molecules connected to cellulose's limited regions often make up cell walls [56]. It is well known that the ripening process of cell walls primarily affects pectin, hemicellulose, and cellulose. This is achieved through the coordinated and cooperative action of enzymes that modify the cell wall. These enzymes primarily consist of polygalacturonase, pectin methylesterase, cellulase, xylanase, β-galactosidase, α-arabinofuranosidase, and protease [57]. Extensive research has been conducted on the action patterns of cell wall modifying enzymes in various fruit varieties, including tomato, pears, zucchini fruit, apples, and so on, in order to investigate the internal relationship between textural properties and fruit softening. Protease enzymes break down proteins into amino acids, contributing also to the softening of fruit texture during ripening [58–61].

4.2 Starch to sugar conversion

The process of fruit ripening also involves the conversion of starch to sugars, which has a substantial impact on the taste and quality of certain fruits, by the fact that amylase enzymes convert starch into sugars, such as glucose and fructose. The development of sugars and organic acids mostly affects the flavor of fruits [62]. In addition, the same authors claim that plastidial β-amylase isoenzyme gene transcript accumulation and enzyme activity were both increased in the later stages of fruit development, suggesting that the enzyme was involved in both the reduction of starch and the rise in total soluble sugar levels in ripe tomatoes.

According to [63], amylases are the most common hydrolase enzymes that break down the glycosidic bonds found in starch molecules to create oligosaccharides and dextrins. Amylases come in two varieties: exo- and endo-amylases. Hydrolyzing the nonreducing end of starch is done by exo-amylases. According to [64], endo-amylases hydrolyze glycosidic bonds inside starch molecules. Amylase is an essential enzyme in biotechnology that is mostly derived from microbes and has a wide range of industrial uses [65].

The interaction between starch modification and other metabolic processes, such as cell wall component disintegration, resulting in the unique texture and flavor profile of ripened produce [66]. The transformation of starch during ripening impacts not only the sensory properties of fruits and vegetables but also their nutritional value. The conversion of starch to sugars adds to a rise in soluble solids, including sugars and organic acids, hence improving the nutritional content of ripe food. The change is significant not only for the fruit's appearance but also for its nutritional value, making it more delicious and accessible for ingestion [67].

4.3 Texture, flavor, and aroma development

Fatty acids are the primary antecedents of flavor in fruits and vegetables and are primarily influenced by the fruits' and vegetables' maturation level, cultivar, region, and processing techniques. The main metabolic process for producing aroma chemicals is β-oxidation, while lipoxygenase is crucial for producing taste compounds from fatty acids. For instance, the β-oxidation pathway links the formation of lactones to the development of the flavors of peaches, nectarine, pineapple (δ-octalactone), coconut (γ-octalactone), and peach (γ-decalactone and γ-dodecalactone, respectively). On the other hand, fruits and vegetables derive their flavor from the lipoxygenase pathway, which breaks down linoleic and linolenic acids into aldehydes, alcohols, acids, and esters [68, 69].

Fruit texture change during ripening is an important characteristic that makes fruit edible, palatable, and appealing for human consumption as well as vectors of fruit dispersal. Textural alterations related to ripening are complex in nature and frequently entail modifications to cell-wall components, including polysaccharides and proteins. Fruit firmness is also affected by elements such as cuticle characteristics, turgor, and free radicals [70].

As previously stated, the ripening of fruits and vegetables is characterized by a wide range of metabolic processes. The action of proteolytic enzymes, which aid in the breakdown of proteins into amino acids, is prominent among them. This enzymatic transformation affects not only the flavor profile of foods but also their nutritional composition. Proteases are enzymes that convert proteins into amino acids. During ripening, this process aids in the softening of tissues and the formation of tastes and smells [71].

The creation of volatile chemicals that give ripe fruits their distinct flavor and aroma depends heavily on enzymes. Enzymes, the molecular developers of biochemical transformations, are critical in the production of volatile chemicals that contribute to the flavor and perfume of ripe fruits. Volatile biosynthetic enzymes, which catalyze the conversion of non-volatile precursors into aroma-rich molecules, are an important group of enzymes engaged in this process.

The enzymatic process of volatile compound production is not merely a biochemical phenomenon; it profoundly influences the sensory experience of consumers. The interplay of various enzymes creates a harmonious blend of volatile chemicals, imparting a signature aroma and flavor to each ripe fruit. This sensory attraction is a result of the complex biochemical movement that occurs within the fruit during the ripening process [72].

Lipid (fat) hydrolysis is catalyzed by lipases. During the ripening phase, this mechanism may play a role in the development of volatile chemicals that give foods their distinctive scents as well as changes in texture and flavor [73].

4.4 Ethylene production and action

ACS (1-Aminocyclopropane-1-carboxylic acid synthase) and ACO (1-Aminocyclopropane-1-carboxylic acid oxidase): these enzymes are involved in the biosynthesis of ethylene, a plant hormone that plays a crucial role in fruit ripening and senescence [74]. The authors mention two specific phases in the ethylene production pathway. S-adenosyl-L-methionine (SAM) is initially metabolized by ACC-synthase (ACS) into 1-aminocyclopropane-1-carboxylic acid (ACC). In a subsequent process, ACC-oxidase (ACO) transforms ACC into ethylene.

Among these changes, the breakdown of chlorophyll is a major step, signifying the transition from the green, unripe stage to the vivid hues associated with ripeness. The main regulator of ripening processes, ethylene, also modulates chlorophyll degradation. Ethylene promotes the production and activation of genes linked with chlorophyll degradation enzymes. As fruits and vegetables mature from unripe to ripe, ethylene orchestrates a coordinated enzymatic reaction, enabling efficient chlorophyll breakdown and the appearance of vibrant hues [1].

Enzymes involved in the decomposition of cell structures, such as peroxidase and polyphenol oxidase, contribute to the breakdown of chlorophyll. As fruits ripen, the cell walls alter, allowing chloroplasts containing chlorophyll to be released. The liberated chloroplasts are then subjected to the enzyme system that breaks down chlorophyll [75]. Enzymatic chlorophyll decomposition improves not only the visual appeal of fruits and vegetables but also has culinary and nutritional ramifications. Color variations are frequently related to change in flavor and nutritional content, impacting customer preferences and perceptions of ripeness and freshness [76].

Fleshy fruits were previously classified as either climacteric or nonclimacteric. At the start of ripening, all climacteric fruits—which include both monocot and dicot species like apple, pear, banana, mango, tomato, and many more—produce an increase in ethylene production and transpiration. It is well established that ethylene and master transcription factors (TFs) with multiple gene targets regulate this "climacteric," which is a precursor to the expression of genes encoding enzymes that catalyze ripening changes. Ripeness requires alterations in gene expression in both climacteric and nonclimacteric fruits [77].

Regarding the tomato fruits among the Solanaceae, α-tomatine is a crucial secondary metabolite. Several genes and enzymes complete its metabolism and biosynthesis. The production and metabolism of α-tomatine are influenced by many transcriptional regulatory factors and hormonal pathways, such as JA, GA, ethylene, light signals, and their downstream transcriptional regulatory factors, in addition to the distinct stages of plant growth and development [78].

4.5 Antioxidant defense

The first-line antioxidant defense system, which is made up of the activities of glutathione peroxidase (GPX), catalase (CAT), and superoxide dismutase (SOD), is essential to the overall defense mechanisms and tactics in biological structures. These enzymes are part of the antioxidant defense system, helping to neutralize reactive oxygen species and preserve the quality of fruits [79].

According to [80, 81], in certain concentrations, fumigation with nitric oxide is helpful in preventing the browning of the surface of freshly cut apple slices. Both the amount of nitric oxide present and the storage temperature have an impact on efficacy. Furthermore, the results suggest that nitric oxide therapy may be used to extend the postharvest shelf life of apple slices across a range of cultivars correlated with lower temperatures.

Regarding the prevention of enzymatic browning another study suggests utilizing an immersion treatment with 0.25% gum and 0.25% sodium chloride to stop enzymatic browning and maintain the quality attributes of pear slices before the drying process is recommended based on the findings of qualitative, structural, and sensory evaluations [82].

Tannins, compounds known as antioxidants, are types of secondary metabolites in plants and enzymes play an important role in their biosynthesis. Condensed tannins, also known as proanthocyanidins, and hydrolyzable tannins are the two main groups of tannins. In addition to their well-known functions in plant defense and human health, both forms of tannins have been suggested as significant molecules for the flavor perception of various fruits and beverages, especially wine [83].

Related to *alkaloids*, modifying enzymes from several families work on a wide variety of alkaloids throughout the kingdom of plants, producing a wide range of physiologically significant alkaloid derivatives. These enzymes catalyze a variety of chemical modification reactions, principally methylation, glycosylation, oxidation, reduction, hydroxylation, and acylation. Certain alkaloids possess antibacterial and antifungal characteristics, as α-tomatin found in tomatoes, piperine found in black pepper, and protoberins and berberins found in a variety of plant species [67].

Fruit handling, storage, and processing after harvest depend on an understanding of and ability to regulate these enzymatic activities. Controlling enzymatic activity can help to prolong fruit shelf life, control ripening, and improve fruit quality for customers.

5. New discoveries in artificial ripening processes

Fruits go through a process called ripening that makes them more palatable. Fruits generally get softer, less green, and sweeter as they ripen. But even though the fruit's sweetness and acidity increase as it ripens, it still tastes sweeter. Fruits are an essential component of the human diet and are very healthy. However, because of their limited shelf life, these are extremely perishable [84]. The natural process of fruit ripening involves the fruit going through a number of chemical changes that eventually cause it to become soft, edible, sweet, and colorful [2, 52]. As science and technology have advanced, different artificial fruit ripening techniques have been noticed, primarily to satisfy customer demand and other commercial factors. When fruit traders sell their products to consumers before the fruit is in season, artificial fruit ripening becomes necessary. Fruit sellers experience a financial loss as a result, thus to reduce that loss, they occasionally choose to gather fruits before they are quite ripe and artificially mature them before selling them to customers. During the off-season, it is simpler to distinguish artificially ripened fruit. Nonetheless, it is more difficult to distinguish between naturally ripened fruits and artificially ripened fruits in terms of appearance during the ripening season. Fruit traders artificially mature green fruits even in the appropriate season in order to satisfy the increased demand and maximize profits from seasonal fruits. In order to address the problems with distribution and transportation, they also artificially ripen fruits [1, 2].

Additionally, regular consumption of artificially ripened fruits has been linked by scientists to heart-related disorders, weakness, dizziness, and skin ulcers [85, 86]. Furthermore, these ripening agents could include various contaminants that are hazardous to human health. A variety of laws and legislation have been made and put into effect in various countries to limit or outright forbid the production, sale, and distribution of artificial fruit ripening agents in response to growing health-related concerns [84, 87].

To determine the relevant health risk, it is crucial to do both qualitative and quantitative investigation of the ripening agents present in the fruit's peel and flesh. On the skin of the fruit, artificial ripening chemicals are typically present. Analyzing the chemical influence of artificially ripened fruits on their food value and quantifying the compounds present in fruit flesh are also crucial [84, 87].

The main ripening chemical that fruits naturally produce and that starts the ripening process is ethylene. Numerous ripening agents have a variety of applications that include the release of ethylene to accelerate the ripening process. Artificial ripening of fruits and vegetables is achieved through the use of chemicals such as ethanol, glycol, methanol, ethylene, ethephon, and calcium carbide [32, 88, 89].

To encourage the ripening of fruit, just 1 ppm of ethylene in the air is needed. Applying ethylene externally has the potential to induce or trigger the natural ripening process of various fruits and vegetables, including avocado, banana, mango, papaya, pineapple, and guava. As a result, these products can be provided before the schedule.

Throughout the world, calcium carbide is frequently used. Although calcium carbide, which is sold as a grayish-black powder, is mostly used in welding, it is also widely used in many developing nations to artificially ripen climacteric fruits. Calcium carbide-ripened fruits have a smooth texture and well-developed peel color, although they lose flavor. Calcium carbide, commonly referred to as masala, is widely used in papayas, bananas, and mangoes, and occasionally in apples and plums. Due to its low cost, dealers utilize it carelessly instead of adhering to other suggested methods of ripening, such as dipping fruits in ethephon solution or subjecting them to ethylene gas. When calcium carbide is applied to fruits, it reacts with moisture to release acetylene, which is similar to ethylene in how it ripens fruit. Traces of arsenic and phosphorus hydride, which are harmful to human health when in direct contact with them, are present in industrial-grade calcium carbide [16, 90].

Another substance that is used to artificially ripen fruits is ethephon. Since pineapple, banana, and tomato treated with 1000 ppm of ethephon required less time to mature (48, 32, and 50 hours, respectively) than other treated fruits as well as compared with the nontreated fruits, ethephon is frequently seen to be preferable than calcium carbide. Researchers from Bangladesh found that artificially ripened pineapples and bananas have lower nutritional values, such as protein content, vitamin C, and beta-carotene. However, the most important discovery was the presence of arsenic (As) and lead (Pb) in these artificially ripened fruits and vegetables. Although the concentrations of As and Pb were within the adult daily permitted intake limit, regular consumption of these fruits can result in major health risks for people, including skin irritation, cancer, liver illness, renal disease, and diarrhea.

Artificial ripening agents ideally should not contain metal or metallicoids and instead release ethylene or acetylene to initiate fruit ripening. However, a large amount of As, Pb, and phosphorus compounds-which are hazardous to human health and can contaminate artificially ripened fruits-may be present in nearly industrial-grade calcium carbide and ethephon. When using high-quality ripening agents, it's important to eliminate metal/metalloid contamination and use modest dosage rates [17, 22, 84].

The ripening process is accelerated by acetylene, which is released by calcium carbide and functions similarly to ethylene. It has been discovered that directly consuming acetylene is harmful since it lowers the brain's oxygen supply and may even result in chronic hypoxia [85]. Because of its alkaline nature, calcium carbide irritates the abdominal mucosal tissue. There have been cases recently of gastric distress following consumption of mangoes ripened by carbide. Although eating fruit that has ripened with carbide does not immediately cause an allergic reaction, applying these chemicals to the fruit may cause dizziness or headaches.

Both established and developing nations have passed and put into effect several statutes and rules limiting the use of artificial fruit ripening agents in order to address the growing health-related concerns. Formally recommended by the National Organic Standards Board (NOSB) to the Organic Program (NOP), ethylene is used in the United States to aid in the postharvest ripening of tropical fruit and the degreening of citrus. Soil Association Organic Standards, version 16.4, June 2011 states that ethylene can be used to ripen bananas and kiwis in the United Kingdom. Additionally, ethylene gas is included on the IFOAM Indicative List of Substances for Organic Production and Processing as being "Only for ripening fruits" by the International Federation of Organic Agriculture Movements (IFOAM).

Commercial ripening is a crucial aspect of the fruit industry since ripe fruits degrade easily and should not be carried or distributed. Thus, fruit dealers select underride fruits and apply specific techniques to prolong their shelf life. It is desirable, in this sense, to use chemical approaches that are both valid and acceptable. Anything that goes against that could be harmful to our health.

Artificial fruit ripening is a complicated problem that calls for the cooperation of government organizations, legislators, fruit suppliers, farmers, scientists, and customers to find a workable solution, particularly in developing nations. To address the problems, it is crucial to evaluate many aspects of artificial fruit ripening, look at accepted procedures, and conduct in-depth scientific research as opposed to generalizing the problem [17, 22, 84, 85, 87].

6. Conclusions

The aim of this chapter was to cover the latest methods used in fruit and vegetable ripening processes, food industry ripening processes, enzymatic ripening processes, and artificial ripening processes.

The most significant characteristics influenced by the ripening of fruits and vegetables are: texture, color, aroma, smell, nutrient metabolism, and other quality characteristics that, ultimately, make the fruit attractive, desirable, and edible for consumers. These processes are greatly influenced by temperature, water, fertilization processes, and pest control, but also new technologies such as genetic engineering, drone technology, robotics, automation based on the Internet of Things, and smartphone applications.

Postharvest processes help to reduce losses through, ethylene control, storage conditions, environmental agents, techniques, commodities, as well as the season and location of production.

The ripening processes of processed foods are dependent on microorganisms, including molds, yeasts, gram-positive catalase-positive cocci (GCC+), and lactic acid bacteria (LAB), which exert a beneficial influence on the formation of desired sensory attributes. Preservation processes include physical techniques such as heat treatments, ionizing radiation, and high hydrostatic pressures, as well as chemical preservatives such as organic acids, antifungal compounds, nitrates, and nitrites.

The best-known enzymatic activities involved in the ripening process are: cell wall degradation, starch conversion into sugar, texture, flavor, and aroma development, ethylene production and action, antioxidant defense, tannins, and alkaloids.

The most commonly used chemicals for the artificial ripening of fruits and vegetables are by using acetylene, ethanol, glycol, methanol, ethylene, ethephon, and calcium carbide.

These processes aim to provide food of superior quality, high textural, and sensory properties, at the highest level of security and safety.

Author details

Romina Alina Marc[1*], Crina Carmen Mureșan[1], Anamaria Pop[1], Georgiana Smaranda Marțiș[1], Andruța Elena Mureșan[1], Alina Narcisa Postolache[2], Florina Stoica[3], Ioana Cristina Crivei[4], Ionuț-Dumitru Veleșcu[5] and Roxana Nicoleta Rațu[5]

1 Faculty of Food Science and Technology, Food Engineering Department, University of Agricultural Science and Veterinary Medicine Cluj-Napoca, Cluj-Napoca, Romania

2 Research and Development Station for Cattle Breeding Dancu, Iasi, Romania

3 Faculty of Agriculture, Department of Pedotechnics, "Ion Ionescu de la Brad" University of Life Sciences, Iasi, Romania

4 Faculty of Veterinary Medicine, Department of Public Health, "Ion Ionescu de la Brad" University of Life Sciences, Iasi, Romania

5 Faculty of Agriculture, Department of Food Technology, "Ion Ionescu de la Brad" University of Life Sciences, Iasi, Romania

*Address all correspondence to: romina.vlaic@usamvcluj.ro

IntechOpen

© 2023 The Author(s). Licensee IntechOpen. This chapter is distributed under the terms of the Creative Commons Attribution License (http://creativecommons.org/licenses/by/3.0), which permits unrestricted use, distribution, and reproduction in any medium, provided the original work is properly cited. (cc) BY

References

[1] Bouzayen M et al. Mechanism of fruit ripening. Plant Developmental Biology - Biotechnological Perspectives. 2010;**1**:319-339

[2] Liu Y et al. Editorial: Advances in ripening regulation, quality formation, pre and post-harvest applications of horticultural products. Frontiers in Plant Science. 2023;**14**

[3] Chang C. Q&A: How do plants respond to ethylene and what is its importance? BMC Biology. 2016;**14**(1):7

[4] Thole V, Vain P, Martin C. Effect of elevated temperature on tomato post-harvest properties. Plants (Basel). 2021;**10**(11):2359

[5] Li S, Chen K, Grierson D. Molecular and hormonal mechanisms regulating fleshy fruit ripening. Cell. 2021;**10**(5):1136

[6] Negri A et al. Proteome changes in the skin of the grape cultivar Barbera among different stages of ripening. BMC Genomics. 2008;**9**:378

[7] Cherian S, Figueroa CR, Nair H. 'Movers and shakers' in the regulation of fruit ripening: A cross-dissection of climacteric versus non-climacteric fruit. Journal of Experimental Botany. 2014;**65**(17):4705-4722

[8] Matias P et al. Citrus pruning in the mediterranean climate: A review. Plants (Basel). 2023;**12**(19):3360

[9] Mészáros M et al. Linking mineral nutrition and fruit quality to growth intensity and crop load in apple. Agronomy. 2021;**11**(3):506

[10] Sidhu RS, Bound SA, Hunt I. Crop load and thinning methods impact yield, nutrient content, fruit quality, and physiological disorders in "scilate". Apples. 2022;**12**(9):1989

[11] Bhavsar D et al. A comprehensive and systematic study in smart drip and sprinkler irrigation systems. Smart Agricultural Technology. 2023;**5**:100303

[12] Jariwala H et al. Controlled release fertilizers (CRFs) for climate-smart agriculture practices: A comprehensive review on release mechanism, materials, methods of preparation, and effect on environmental parameters. Environmental Science and Pollution Research. 2022;**29**(36):53967-53995

[13] Brunner JF. Integrated pest management in tree fruit crops. Food Reviews International. 2014;**2**:15-30

[14] Balasundram SK et al. The role of digital agriculture in mitigating climate change and ensuring food security: An overview. Sustainability. 2023;**15**(6):5325

[15] Strano MC et al. Postharvest technologies of fresh citrus fruit: Advances and recent developments for the loss reduction during handling and storage. Progress in Nutrition. 2022;**8**(7):612

[16] Pott DM, Vallarino JG, Osorio S. Metabolite changes during postharvest storage: Effects on fruit quality traits. Metabolites. 2020;**10**(5):187

[17] Bai L, Liu M, Sun Y. Overview of food preservation and traceability technology in the smart cold chain system. Food. 2023;**12**(15):2881

[18] Mahajan PV et al. Postharvest treatments of fresh produce. Philosophical Transactions of the Royal Society A. 2017;**2014**(372):20130309

[19] Mariah MAA et al. The emergence and impact of ethylene scavengers techniques in delaying the ripening of fruits and vegetables. Membranes (Basel). 2022;**12**(2):117

[20] Calín-Sánchez Á et al. Comparison of traditional and novel drying techniques and its effect on quality of fruits. Vegetables and Aromatic Herbs. 2020;**9**(9):1261

[21] Dawson P, Al-Jeddawi W, Rieck J. The effect of different freezing rates and long-term storage temperatures on the stability of sliced peaches. International Journal of Food Science. 2020;**2020**:9178583

[22] Fang Y, Wakisaka M. A review on the modified atmosphere preservation of fruits and vegetables with cutting-edge technologies. Agriculture. 2021;**11**(10):992

[23] Matloob A et al. A review on edible coatings and films: Advances, composition, production methods, and safety concerns. ACS Omega. 2023;**8**(32):28932-28944

[24] Alam AU et al. Fruit quality monitoring with smart packaging. Sensors (Basels). 2021;**21**(4):1509

[25] Rop O et al. Effect of five different stages of ripening on chemical compounds in medlar (Mespilus germanica L.). Molecules. 2010;**16**(1):74-91

[26] Lers A. 27 - Potential application of biotechnology to maintain fresh produce postharvest quality and reduce losses during storage. In: Altman A, Hasegawa PM, editors. Plant Biotechnology and Agriculture. San Diego: Academic Press; 2012. pp. 425-441

[27] Carvalho DUd et al. Effectiveness of natural-based coatings on sweet oranges post-harvest life and antioxidant capacity of obtained by-products. Horticulturae. 2023;**9**(6):635

[28] Alós E, Rodrigo MJ, Zacarias L. Chapter 7 - Ripening and Senescence. In: Yahia EM, editor. Postharvest Physiology and Biochemistry of Fruits and Vegetables. Oxford, UK: Woodhead Publishing; 2019. pp. 131-155

[29] Wilhelm C. Encyclopedia of applied plant sciences. Journal of Plant Physiology. 2004;**161**:1186-1187

[30] Klee HJ, Giovannoni JJ. Genetics and control of tomato fruit ripening and quality attributes. Annual Review of Genetics. 2011;**45**:41-59

[31] Seefeldt HF, Løkke MM, Edelenbos M. Effect of variety and harvest time on respiration rate of broccoli florets and wild rocket salad using a novel O2 sensor. Postharvest Biology and Technology. 2012;**69**:7-14

[32] Alonso-Salinas R et al. Effect of potassium permanganate, ultraviolet radiation and titanium oxide as ethylene scavengers on preservation of postharvest quality and sensory attributes of broccoli stored with tomatoes. Foods. 2023;**12**(12):2418

[33] Kou J et al. Effects of ethylene and 1-methylcyclopropene on the quality of sweet potato roots during storage: A review. Horticulturae. 2023;**9**(6):667

[34] López-Gómez A, Navarro-Martínez A, Martínez-Hernández GB. Active paper sheets including nanoencapsulated essential oils: A green packaging technique to control ethylene production and maintain quality in fresh horticultural products-a case study on flat peaches. Food. 2020;**9**(12):1904

[35] Nybom H et al. Review of the impact of apple fruit ripening, texture

and chemical contents on genetically determined susceptibility to storage rots. Plants (Basel). 2020;**9**(7):831

[36] Wills RBH. Potential for more sustainable energy usage in the postharvest handling of horticultural produce through management of ethylene. Journal of the Science of Food and Agriculture. 2021;**9**(10):147

[37] Janská A et al. Cold stress and acclimation - What is important for metabolic adjustment? Plant Biology (Stuttgart, Germany). 2010;**12**:395-405

[38] Delgado J et al. Biocontrol of pathogen microorganisms in ripened foods of animal origin. Microorganisms. 2023;**11**(6):1578

[39] Camargo A et al. Microbiological quality and safety of Brazilian artisanal cheeses. Brazilian Journal of Microbiology. 2021;**52**:393-409

[40] Martín I et al. Growth and expression of virulence genes of Listeria monocytogenes during the processing of dry-cured fermented "salchichón" manufactured with a selected lactilactobacillus sakei. Biology (Basel). 2021;**10**(12):1258

[41] Asensio M et al. Control of toxigenic molds in food processing. Microbial Food Safety and Preservation Techniques. 2014;**1**:329-358

[42] Balasubramaniam VM, Lee J, Serventi L. Understanding new foods: Development of next generation of food processing, packaging, and ingredients technologies for clean label foods. In: Serventi L, editor. Sustainable Food Innovation. Cham: Springer International Publishing; 2023. pp. 157-167

[43] Forde A, Fitzgerald GF. Biotechnological approaches to the understanding and improvement of mature cheese flavour. Current Opinion in Biotechnology. 2000;**11**(5):484-489

[44] Cocolin L et al. Next generation microbiological risk assessment meta-omics: The next need for integration. International Journal of Food Microbiology. 2018;**287**:10-17

[45] Fox PF et al. Dairy Chemistry and Biochemistry. Second ed. Switzerland: Springer; 2015. pp. 1-584

[46] Mureşan CC et al. Changes in physicochemical and microbiological properties, fatty acid and volatile compound profiles of apuseni cheese during ripening. Food. 2021;**10**(2):258

[47] Khattab AR et al. Cheese ripening: A review on modern technologies towards flavor enhancement, process acceleration and improved quality assessment. Trends in Food Science & Technology. 2019;**88**:343-360

[48] Smit G, Smit BA, Engels WJM. Flavour formation by lactic acid bacteria and biochemical flavour profiling of cheese products. FEMS Microbiology Reviews. 2005;**29**(3):591-610

[49] Wang D et al. The changes occurring in proteins during processing and storage of fermented meat products and their regulation by lactic acid bacteria. Foods. 2022;**11**(16):2427

[50] Tatiyaborworntham N et al. Paradoxical effects of lipolysis on the lipid oxidation in meat and meat products. Food Chemistry: X. 2022;**14**:100317

[51] Symons GM et al. Hormonal changes during non-climacteric ripening in strawberry. Journal of Experimental Botany. 2012;**63**(13):4741-4750

[52] Prasanna V, Prabha TN, Tharanathan RN. Fruit ripening

phenomena - An overview. Critical Reviews in Food Science and Nutrition. 2007;**47**(1):1-19

[53] Martínez-Romero D et al. Tools to maintain postharvest fruit and vegetable quality through the inhibition of ethylene action: A review. Critical Reviews in Food Science and Nutrition. 2007;**47**(6):543-560

[54] Paniagua C et al. Fruit softening and pectin disassembly: An overview of nanostructural pectin modifications assessed by atomic force microscopy. Annals of Botany. 2014;**114**(6):1375-1383

[55] Goulao LF, Oliveira CM. Cell wall modifications during fruit ripening: When a fruit is not the fruit. Trends in Food Science & Technology. 2008;**19**(1):4-25

[56] Ren Y-Y et al. Degradation of cell wall polysaccharides and change of related enzyme activities with fruit softening in Annona squamosa during storage. Postharvest Biology and Technology. 2020;**166**:111203

[57] Barka EA et al. Impact of UV-C irradiation on the cell wall-degrading enzymes during ripening of tomato (Lycopersicon esculentum L.) fruit. Journal of Agricultural and Food Chemistry. 2000;**48**(3):667-671

[58] Carvajal F et al. Cell wall metabolism and chilling injury during postharvest cold storage in zucchini fruit. Postharvest Biology and Technology. 2015;**108**:68-77

[59] Wei J et al. Changes and postharvest regulation of activity and gene expression of enzymes related to cell wall degradation in ripening apple fruit. Postharvest Biology and Technology. 2010;**56**(2):147-154

[60] Dong Y, Zhang S, Wang Y. Compositional changes in cell wall polyuronides and enzyme activities associated with melting/mealy textural property during ripening following long-term storage of 'Comice' and 'd'Anjou' pears. Postharvest Biology and Technology. 2018;**135**:131-140

[61] Bu J et al. Postharvest UV-C irradiation inhibits the production of ethylene and the activity of cell wall-degrading enzymes during softening of tomato (Lycopersicon esculentum L.) fruit. Postharvest Biology and Technology. 2013;**86**:337-345

[62] Maria T et al. Gene transcript accumulation and enzyme activity of β-amylases suggest involvement in the starch depletion during the ripening of cherry tomatoes. Plant Gene. 2016;**5**:8-12

[63] Sundarram A, Murthy TPK. α-Amylase production and applications: A review. Journal of Applied & Environmental Microbiology. 2014;**2**:166-175

[64] Kamon M et al. Characterization and gene cloning of a maltotriose-forming exo-amylase from Kitasatospora sp. MK-1785. Applied Microbiology and Biotechnology. 2015;**99**(11):4743-4753

[65] Farooq MA et al. Biosynthesis and industrial applications of alpha-amylase: A review. Archives of Microbiology. 2021;**203**(4):1281-1292

[66] Baldwin L et al. Structural alteration of cell wall pectins accompanies pea development in response to cold. Phytochemistry. 2014;**104**:37-47

[67] Bhambhani S, Kondhare KR, Giri AP. Diversity in chemical structures and biological properties of plant alkaloids. Molecules. 2021;**26**(11):3374

[68] Distefano M et al. Aroma volatiles in tomato fruits: The role of genetic,

preharvest and postharvest factors. Agronomy. 2022;**12**(2):376

[69] Shahidi F, Hossain A. Role of lipids in food flavor generation. Molecules. 2022;**27**(15):5014

[70] Chaïb J et al. Physiological relationships among physical, sensory, and morphological attributes of texture in tomato fruits. Journal of Experimental Botany. 2007;**58**(8):1915-1925

[71] Perotti VE, Moreno AS, Podestá FE. Physiological aspects of fruit ripening: The mitochondrial connection. Mitochondrion. 2014;**17**:1-6

[72] Amos RA, Mohnen D. Critical review of plant cell wall matrix polysaccharide glycosyltransferase activities verified by heterologous protein expression. Frontiers in Plant Science. 2019;**10**:915

[73] Panzanaro S et al. Biochemical characterization of a lipase from olive fruit (Olea europaea L.). Plant Physiology and Biochemistry. 2010;**48**(9):741-745

[74] Houben M, Van de Poel B. 1-Aminocyclopropane-1-carboxylic acid oxidase (ACO): The enzyme that makes the plant hormone ethylene. Frontiers in Plant Science. 2019;**10**:695

[75] Vicente A et al. The linkage between cell wall metabolism and fruit softening: Looking to the future. Journal of the Science of Food and Agriculture. 2007;**87**:1435-1448

[76] Goff SA, Klee HJ. Plant volatile compounds: Sensory cues for health and nutritional value? Science. 2006;**311**(5762):815-819

[77] Li S, Chen K, Grierson D. A critical evaluation of the role of ethylene and MADS transcription factors in the network controlling fleshy fruit ripening. The New Phytologist. 2019;**221**(4):1724-1741

[78] Liu Y et al. Current advances in the biosynthesis, metabolism, and transcriptional regulation of alpha-tomatine in tomato. Plants (Basel). 2023;**12**(18):3289

[79] Ighodaro OM, Adeosun AM, Akinloye OA. Alloxan-induced diabetes, a common model for evaluating the glycemic-control potential of therapeutic compounds and plants extracts in experimental studies. Medicina (Kaunas, Lithuania). 2017;**53**(6):365-374

[80] Pristijono P, Wills RBH, Golding JB. Inhibition of browning on the surface of apple slices by short term exposure to nitric oxide (NO) gas. Postharvest Biology and Technology. 2006;**42**(3):256-259

[81] Huque R et al. Effect of nitric oxide (NO) and associated control treatments on the metabolism of fresh-cut apple slices in relation to development of surface browning. Postharvest Biology and Technology. 2013;**78**:16-23

[82] Alipoorfard F, Jouki M, Tavakolipour H. Application of sodium chloride and quince seed gum pretreatments to prevent enzymatic browning, loss of texture and antioxidant activity of freeze dried pear slices. Journal of Food Science and Technology. 2020;**57**(9):3165-3175

[83] Mora J et al. Regulation of plant tannin synthesis in crop species. Frontiers in Genetics. 2022;**13**:870976

[84] Gandhi S, Sharma M, Bhatnagar B. Comparative study on the ripening ability of artificial ripening agent (Calcium Carbide) and natural ripening agents. Global Journal of Biology,

Agriculture and Health Science. 2019;**5**(2):106-110

[85] Fattah SA, Ali MY. Carbide ripened fruits-A recent health hazard. Faridpur Medical College Journal. 2011;5(2)

[86] Gross KC et al. Biochemical changes associated with the ripening of hot pepper fruit. Physiologia Plantarum. 1986;**66**:31-36

[87] Mursalat M et al. A critical analysis of artificial fruit ripening: scientific, legislative and socio-economic aspects. Food Technology. 2013;**04**:6-12

[88] Goonatilake, R. Effects of Diluted Ethylene Glycol as A Fruit-Ripening Agent. Global Journal of Biotechnology and Biochemistry. IDOSI Publications; 2008;**3**(1):08-13. ISSN 2078-466X

[89] Ruwali A et al. Effect of different ripening agents in storage life of banana (Musa paradisiaca) at Deukhuri, Dang, Nepal. Journal of Agriculture and Food Research. 2022;**10**:100416

[90] Per H et al. Calcium carbide poisoning via food in childhood. The Journal of Emergency Medicine. 2007;**32**(2):179-180

Chapter 2

The First Signal to Initiate Fruit Ripening is Generated in the Cuticle: An Hypothesis

Miguel-Angel Hernández-Oñate,
Eduardo-Antonio Trillo-Hernández
and Martín-Ernesto Tiznado-Hernández

Abstract

The paradigm that has prevailed for a long time sustains that ethylene is the first signal that initiates fruit ripening. However, in this manuscript, we present the hypothesis that a signal generated from the cuticle induces the synthesis of ethylene, and therefore, it is the initial signal that triggers the fruit-ripening phenomena. Among the experimental evidence supporting the hypothesis, we can mention that cuticle components released during the plant pathogenic attack can induce the synthesis of ethylene in plants. Also, it has been found that in fungi, a cuticle component can activate a transcription factor by phosphorylation, which induces the transcription of a gene encoding a cutinase. Besides, studies with plant tissues experiencing a high rate of cell expansion have shown that there is a careful synchronization between the demand of cuticle components and biosynthesis, which suggests that the plant cell can sense the moment in which the fruit would stop growing by cell expansion, and initiate the ripening phenomena. In this chapter, experimental evidences supporting the physiological role of the fruit cuticle in the fruit ripening phenomena will be presented and reviewed with the goal to show a possible role of the fruit cuticle in the onset of fruit ripening.

Keywords: cuticle, components, signal, fruit, ripening

1. Introduction

The focus of the chapter is to show the physiological role of the fruit cuticle changes mentioned during fruit ripening. The different reports in which the changes in gene expression related with the biosynthesis, transport, and assembly of the different cuticle components during fruit development will not be discussed, unless they are relevant to support the hypothesis of this work. Readers interested in this subject can consult several excellent reviews [1–3].

Cuticle is a thin layer made of large molecular weight molecules with hydrophobic characteristics as well as polysaccharides and phenolic compounds. Its hydrophobic physical properties make the diffusion of non-lipophilic molecules such as water or other charged molecules very difficult [4]. In this way, the cuticle plays a very important role of reducing the water loss in the surrounding environment, which is one of the cuticle's most important physiological functions. Besides that, the cuticle plays a role as a defense of the fruit against different biotic and abiotic stresses such as fungi and bacteria infection, heat, mechanical support, responses to mechanical injury, responses to heat, protection against the negative effects of UV light, plant development, as well as physiological disorders such as fruit cracking [5–9].

The role of ethylene as a signal to begin fruit ripening is very old knowledge. Indeed, there are many experimental evidence in support that ethylene plays an important role in inducing the expression of genes, which carries out the different changes that the fruit shows during fruit ripening [10, 11]. However, it is important to challenge this idea in pursuit of a signal, which can begin the synthesis of ethylene. Based on the above mentioned, it is clear that the cuticle plays an important role in several fruit physiological phenomena. Further, the cuticle composition is very complex, and the possibility of a signal enzymatically released from the cuticle components when the fruit has finished growing by cell expansion can in turn induce the biosynthesis of ethylene to start the fruit ripening. In fact, it has been suggested that cuticle can control the fruit growth rate. Alternatively, it is possible to think a change in cuticle biomechanics can be sensed by a receptor to start the fruit ripening.

The main goal of the present chapter is to provide evidence supporting the hypothesis that the cuticle is playing an important role in the onset of fruit ripening.

2. Changes in the bio-mechanical properties of the cuticle during fruit ripening

The study of the biomechanical properties has suggested that the cuticle shows several characteristics that allows to conclude that it is similar to the ones present in smart materials [12]. Comparing the mechanical properties of tomato fruit cuticle with leaf cuticles from several species, it is possible to observe that the tomato fruit cuticle shows a higher tendency to deform as compared with leaf, as concluded from the much lower Young's moduli of the tomato fruit cuticle. This most likely is related to the need of the fruit cuticle to yield under the stress of a lower force because the fruit must increase the size without cracking [13]. Moreover, an increase in the Young's moduli from mature green to red ripe tomato fruits had been observed in several tomato cultivars [14].

Besides the above mentioned, it was found that the cuticle physical properties do not have any correlation with the thickness [13]. Comparison between the cuticle of non-ripening tomato mutants such as alcobaca, rin, and nor with the normal ripening variety of tomato fruit 'Ailsa Craig' found that there was a difference in cuticle lipid composition [15]. Specifically, a larger amount of trihydroxy C18 fatty acids was found during the stages of small green, mature green, and red ripe equivalent in the non-ripening mutants such as alcobaca, rin, and nor as compared with the 'Ailsa Craig' (AC) tomato. The authors suggested that the presence of more C18 compounds could be related with a more elastic cuticle [15]. On the other hand, by studying tomato fruit varieties with large or low amount of flavonoids, it was found that the amount of flavonoids in the cuticle induces an increase in the Young's moduli [14]. In agreement, the inhibition of the chalcone synthase gene by virus inducing gene silencing in two tomato cultivars induced a lower amount of flavonoids and a decrease in the Young's moduli as a consequence [16]. Besides, the viscoelastic properties of the tomato fruit cuticle during the last stages of tomato fruit ripening has been ascribed to the increased amount of phenolics, which interacts with cutin and the different polysaccharide compounds [17, 18]. On the other hand, it has been mentioned that the whole cuticle is behind the elastic properties of cuticle, whereas the viscoelastic behavior of tomato cuticle is related to the cutin component [19]. Wax is related positively to the level of the Young's moduli or stiffness as it was concluded from experiments in which the wax was removed from the cuticle of several tissues and tomato fruit [20]. In contrast, studies carried out in tomato wild type and a mutant lacking the enzyme beta-ketoacyl-coenzyme A synthase, which has low amount of wax in the cuticle, showed that cuticle from normal tomato showed a lower elastic modulus. Further, this characteristic was only observed in mature green tomatoes because in red ripe tomatoes, both the mutant and the wild type showed the same level of stiffness [21]. Studies focused on the changes of the cuticle biomechanical characteristics during tomato fruit ripening allowed the authors to suggest that the cuticle can be controlling the fruit growth rate during ripening, which in turn suggests that the cuticle can control the initiation of fruit growth to start the fruit development and ripening phenomena [22].

Studies in tomato fruit in which the invertase gene was inhibited found large changes in cuticular components and an increase in the cuticle thickness. Further, the data mentioned above suggest that the cell metabolism is closely related to the cuticle biosynthesis, which in turn suggests an active communication mechanism between cells and cuticle behavior. The growth behavior for the transgenic and wild type was similar, which rules out the fact that there was an alteration in the cell expansion, which could have had effects on cuticle accumulation. In turn, this clearly suggests that even though the cuticle is outside the cell, there is a mechanism of data exchange with the plant epidermal cell [23].

From the data included in this section of the chapter, the different components of the fruit cuticle give the particular biomechanical properties of the cuticle. In this regard, it is possible to put forward a hypothetical question about a signal created by changes in the cuticle biomechanics to induce the onset of the fruit ripening. Maybe there is a receptor that can sense these biomechanical cuticle changes, similar to the Arabidopsis receptor located at the cell membrane, known as THESEUS1, which can sense the integrity of the cell wall [24].

3. Involvement of cuticle in the onset of fruit ripening

Experimental evidence supporting the physiological function of ethylene in the initiation of fruit ripening is vast, and several excellent reviews are available on the subject [10, 11, 25]. However, until now, another suggestion about what the first signal initiating fruit ripening is has not been elaborated, to our knowledge.

The cuticle is the outermost layer that covers the fruit. Further, it is made of large molecules of lipophilic nature such as waxes, fatty alcohols, alkanes, and fatty acids [2].

Analysis carried out in several fruits clearly shows that the cuticle changes in amount and composition during fruit growth and development. This had been recorded in tomato [26], pitahaya [27], peach [28], olive [29], mango [30], and many other fruits. The molecular mechanism of cuticle biosynthesis had been studied in tomato [31], mango [30], and sweet cherry [32]. Further, important advances in that mechanism in pitaya fruit (*Stenocereus thurberi*) have been achieved in our laboratory (García-Coronado, unpublished). The studies mentioned above clearly show that there is rather substantial information about the genes playing a role in cuticle biosynthesis. Besides that, there are several transcriptomes of fruits generated by the utilization of the next generation sequencing technology, including chili pepper [33], strawberry fruit [34], avocado [35], litchi [36], and many others. With all this information, an interesting review had been published about the changes in fruit cuticle composition and the molecular biology of cuticle biosynthesis in fleshy fruits [2]. All the scientific information mentioned above clearly shows that there is enough data to test experimentally the present hypothesis.

There are many differences in the cuticle composition in a tomato mutant designated as delayed fruit deterioration, which shows a large postharvest shelf life, as compared with a tomato fruit with a normal shelf life [37]. These data suggest that the fruit cuticle plays a role in the postharvest shelf life of fruits as it had been suggested by other authors [38]. Also, fruit softening is a characteristic associated with the fruit ripening phenomena. In this regard, it had been found that the cuticle plays an important role in tomato fruit softening [39], in contrast with earlier hypotheses stating that it is the plant cell wall that plays the main role during the fruit softening phenomena.

Evidence generated by studying rapidly expanding stems of Arabidopsis demonstrated that the amount and composition of the cuticle are maintained during these phenomena by synchronization between area expansion and the biosynthesis and exportation of cuticle components. Further, it was shown that the rate of expansion in the apical meristem was 10 times faster than in the middle stem section, and still, the cuticle amount and composition did not change [40]. These observations suggest that it is possible for the epidermal cell to detect the demand for cuticle components during the expansion of the plant stem. Likewise, it can be suggested that the fruit epidermis can also perceive when the fruit stops growing by cell expansion, which is the moment when the fruit starts the ripening phenomenon [41].

By doing a search in the Solgenomics database with the software Blastp using a cutinase orthologous gene from Arabidopsis AT4630140, it was possible to find genes that encodes for cutinases in the tomato genome (**Figure 1**). By analyzing **Figure 1**, it is clear that some of the cutinase-encoding genes are expressed before the initiation of fruit ripening. Furthermore, promoter analysis with the software plant promoter (PlantPAN; http://PlantPAN.itps.ncku.edu.tw/) of the first five genes, which shows the highest homology, the presence of the CREB response element in the genes Solyc02g071610, Solyc02g071720, Solyc03g005900 was found. Also, for the two first genes, the CREB element was found within 400 bp upstream of the translation start site. Furthermore, the gene Solyc02g071610 shows a large activity at 10 days postanthesis (DPA), and it is still active at 15 DPA in the outer epidermis. Also, the gene Solyc02g071720 shows a low activity at 5 and 10 DPA. These stages of development are before the initiation of the tomato fruit ripening. Based on the above, it is possible to suggest that the release of the cuticle components can be carried out by any of these genes to initiate the signal to induce the active synthesis of ethylene.

The possible molecular mechanism explaining how a cuticle component can induce the activation of the fruit cutinases and in turn the release of cuticle components was generated with studies in fungi. In those studies, it was found that a component from the cuticle can activate a fungi transcription factor by phosphorylation, which in turn can induce the transcription of a gene encoding cutinase by binding to the CREB (cAMP response element binding) and Sp1 responsive elements

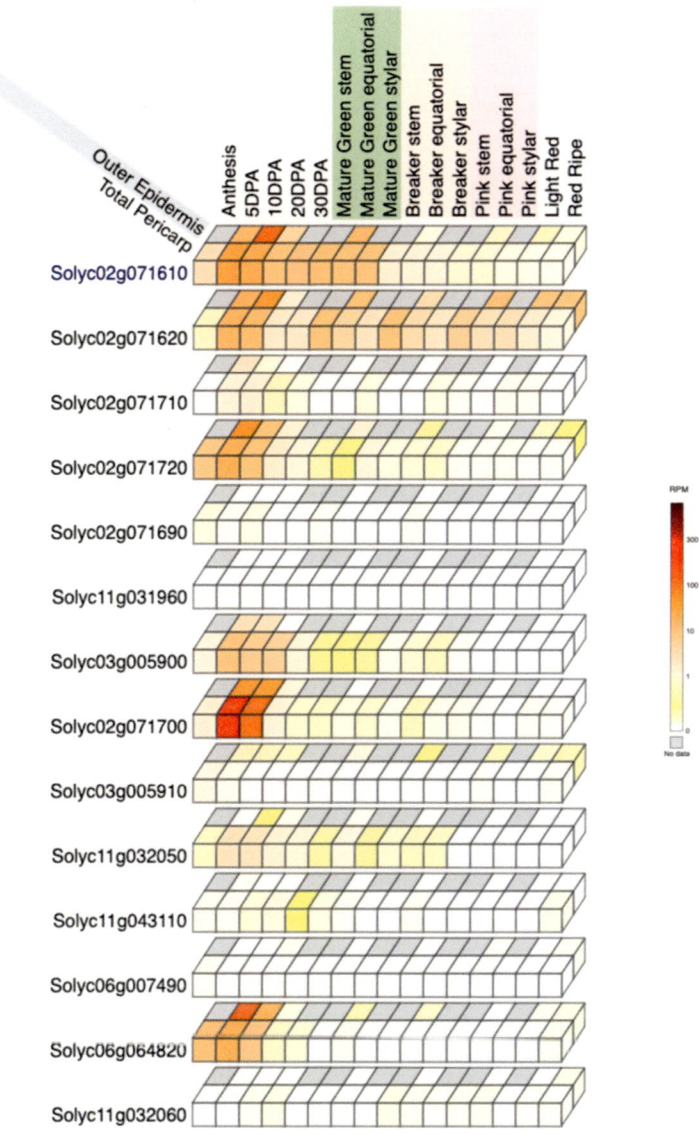

Figure 1.
Changes in expression of tomato genes encoding for Cutinases during the tomato fruit growth and development phenomena. DPA stands for days post Anthesis.

present in the promoter of the gene. Further, the removal of the CREB responsive element eliminates the induction of the gene transcription activation [42]. Also, this responsive element had been found in the 5′-flanking region of the cutinase gene from Colletotrichum gloesporioides and Colletotrichum capsici [43], suggesting that these responsive elements play an important role in the cutinase gene induction. Furthermore, it was found that cycloheximide inhibits the increase in cutinase activity, clearly demonstrating that protein synthesis is involved in the molecular mechanism [44]. Besides the above, it had been found that during ripening, the total number of cutin ester bonds is reduced, making easier the release of cuticle components [45].

On another side, it had been found that the fruit response mechanism in the presence of cuticle components during a pathogen attack includes the production of ethylene, suggesting that there is a mechanism underlying the induction of ethylene biosynthesis by cuticle components [46]. Unfortunately, the molecular mechanism by which a cuticle component can be recognized by the plant cell is not yet known [47], although the role of a receptor had been suggested [48]. It is important to mention that the hypothesis that a mechanism is able to sense a cuticle component by the plant cell is rather old.

Based on the above, our working hypothesis states that a similar mechanism found in fungi can exist in fruits. This in turn implies that a cuticle component released from the cuticle can induce the activation of a transcription factor, which can activate fruit cutinases in the nucleus, which will release cutin components to induce the synthesis of ethylene.

4. Main hypothesis drawbacks

It is important to mention that there are several statements in the hypothesis mentioned that need to be experimentally demonstrated, which are described next: the mechanism for the epidermal cell to detect the moment in which the fruit tissue stops growing by cell expansion; the mechanism explaining the ethylene synthesis induction by a cuticle component; and also, the mechanism by which the plant cell can recognize the cuticle component and the subsequent activation of the gene in the plant nucleus. However, as mentioned above, there is a big knowledge generated by the next-generation sequencing technology about the molecular biology of different fruit physiological phenomena as well as the molecular mechanism of cuticle biosynthesis. Therefore, it is quite possible to carry out experiments to help in the elucidation of the hypothesis elaborated in this document.

5. Conclusion

In this work, a hypothesis has been created as an alternative to explain the mechanism that controls the initiation of fruit ripening. This hypothesis sustains that the first signal to initiate fruit ripening is generated from a cuticle component and not by the hormone ethylene. The evidence supporting the hypothesis is experimental but was not generated by studying fruits. Further, this hypothesis still needs a great deal of experimental evidence to be demonstrated. Also, we believe it is important to challenge the actual paradigm about the fruit ripening phenomena with the recently generated scientific knowledge about the fruit cuticle, which plays roles in several plant physiological phenomena.

Acknowledgements

The authors of the present chapter wish to acknowledge the financial support of Consejo Nacional de Humanidades, Ciencias, y Tecnología of México (CONAHCYT) to carry out this work.

The First Signal to Initiate Fruit Ripening is Generated in the Cuticle: An Hypothesis
DOI: http://dx.doi.org/10.5772/intechopen.112938

Author details

Miguel-Angel Hernández-Oñate[1], Eduardo-Antonio Trillo-Hernández[2,3] and Martín-Ernesto Tiznado-Hernández[4*]

1 CONACYT - Coordinación de Tecnología de Alimentos de Origen Vegetal, Centro de Investigación en Alimentación y Desarrollo A.C. Carretera Gustavo Enrique Astiazarán Rosas, Hermosillo, Sonora, Mexico

2 Unidad de Tecnología de Alimentos-Secretaría de Investigación y Posgrado, Universidad Autónoma de Nayarit, Ciudad de la Cultura S/N, Tepic, Nayarit, Mexico

3 Estancias Posdoctorales-Consejo Nacional de Humanidades, Ciencia y Tecnología, Coordinación de Apoyos a Becarios e Investigadores, Dirección de Posgrado, Ciudad de México, México

4 Coordinación de Tecnología de Alimentos de Origen Vegetal, Centro deInvestigación en Alimentación y Desarrollo A.C. Carretera Gustavo Enrique Astiazarán Rosas, Hermosillo, Sonora, México

*Address all correspondence to: tiznado@ciad.mx

IntechOpen

© 2023 The Author(s). Licensee IntechOpen. This chapter is distributed under the terms of the Creative Commons Attribution License (http://creativecommons.org/licenses/by/3.0), which permits unrestricted use, distribution, and reproduction in any medium, provided the original work is properly cited.

References

[1] Brummell DA, Bowen JK, Gapper NE. Biotechnological approaches for controlling postharvest fruit softening. Current Opinion in Biotechnology. 2022;**78**:102786

[2] Garcia-Coronado H, Tafolla-Arellano JC, Hernandez-Onate MA, Burgara-Estrella AJ, Robles-Parra JM, Tiznado-Hernandez ME. Molecular biology, composition and physiological functions of cuticle lipids in fleshy fruits. Plants-Basel. 2022;**11**(9):1133

[3] Santos M, Egea-Cortines M, Goncalves B, Matos M. Molecular mechanisms involved in fruit cracking: A review. Frontiers in Plant Science. 2023;**14**:1130857

[4] Santier S, Chamel A. Reassessment of the role of cuticular waxes in the transfer of organic molecules through plant cuticles. Plant Physiology and Biochemistry. 1998;**36**(3):225-231

[5] Reynoud N, Petit J, Bres C, Lahaye M, Rothan C, Marion D, et al. The complex architecture of plant cuticles and its relation to multiple biological functions. Frontiers in Plant Science. 2021;**12**:782773

[6] Tafolla-Arellano JC, Gonzalez-Leon A, Tiznado-Hernandez ME, Garcia LZ, Baez-Sanudo R. Composition, physiology and biosynthesis of plant cuticle. Revista Fitotecnia Mexicana. 2013;**36**(1):3-12

[7] Benitez JJ, Moreno AG, Guzman-Puyol S, Heredia-Guerrero JA, Heredia A, Dominguez E. The response of tomato fruit cuticle membranes against heat and light. Frontiers in Plant Science. 2022;**12**:807723

[8] Dominguez E, Fernandez MD, Hernandez JCL, Parra JP, Espana L, Heredia A, et al. Tomato fruit continues growing while ripening, affecting cuticle properties and cracking. Physiologia Plantarum. 2012;**146**(4):473-486

[9] Moreno AG, de Cozar A, Prieto P, Dominguez E, Heredia A. Radiationless mechanism of UV deactivation by cuticle phenolics in plants. Nature Communications. 2022;**13**(1):1786

[10] Li S, Chen K, Grierson D. Molecular and hormonal mechanisms regulating fleshy fruit ripening. Cell. 2021;**10**(5):1136

[11] Kou X, Feng Y, Yuan S, Zhao X, Wu C, Wang C, et al. Different regulatory mechanisms of plant hormones in the ripening of climacteric and non-climacteric fruits: A review. Plant Molecular Biology. 2021;**107**:1-21

[12] Bargel H, Koch K, Cerman Z, Evans NC. Review No. 3Structure–function relationships of the plant cuticle and cuticular waxes—A smart material? Functional Plant Biology. 2006;**33**(10):893-910

[13] Wiedemann P, Neinhuis C. Biomechanics of isolated plant cuticles. Botanica Acta. 1998;**111**(1):28-34

[14] Domínguez E, España L, López-Casado G, Cuartero J, Heredia A. Biomechanics of isolated tomato (Solanum lycopersicum) fruit cuticles during ripening: The role of flavonoids. Functional Plant Biology. 2009;**36**(7):613-620

[15] Kosma DK, Parsons EP, Isaacson T, Lü S, Rose JK, Jenks MA. Fruit cuticle lipid composition during development in tomato ripening mutants. Physiologia Plantarum. 2010;**139**(1):107-117

[16] España L, Heredia-Guerrero JA, Reina-Pinto JJ, Fernández-Muñoz R, Heredia A, Domínguez E. Transient silencing of chalcone synthase during fruit ripening modifies tomato epidermal cells and cuticle properties. Plant Physiology. 2014;**166**(3):1371-1386

[17] Benitez JJ, Guzman-Puyol S, Vilaplana F, Heredia-Guerrero JA, Dominguez E, Heredia A. Mechanical performances of isolated cuticles along tomato fruit growth and ripening. Frontiers in Plant Science. 2021;**12**:787839

[18] González Moreno A, Domínguez E, Mayer K, Xiao N, Bock P, Heredia A, et al. 3D (xyt) Raman imaging of tomato fruit cuticle: Microchemistry during development. Plant Physiology. 2023;**191**(1):219-232

[19] Lopez-Casado G, Matas AJ, Dominguez E, Cuartero J, Heredia A. Biomechanics of isolated tomato (Solanum lycopersicum L.) fruit cuticles: The role of the cutin matrix and polysaccharides. Journal of Experimental Botany. 2007;**58**(14):3875-3883

[20] Khanal BP, Grimm E, Finger S, Blume A, Knoche M. Intracuticular wax fixes and restricts strain in leaf and fruit cuticles. New Phytologist. 2013;**200**(1):134-143

[21] Ehret DL, Frey B, Helmer T, Aharoni A, Wang ZH, Jetter R. Fruit cuticular and agronomic characteristics of a lecer6 mutant of tomato. Journal of Horticultural Science & Biotechnology. 2012;**87**(6):619-625

[22] Bargel H, Neinhuis C. Tomato (Lycopersicon esculentum mill.) fruit growth and ripening as related to the biomechanical properties of fruit skin and isolated cuticle. Journal of Experimental Botany. 2005;**56**(413):1049-1060

[23] Vallarino JG, Yeats TH, Maximova E, Rose JK, Fernie AR, Osorio S. Postharvest changes in LIN5-down-regulated plants suggest a role for sugar deficiency in cuticle metabolism during ripening. Phytochemistry. 2017;**142**:11-20

[24] Hématy K, Sado PE, Van Tuinen A, Rochange S, Desnos T, Balzergue S, et al. A receptor-like kinase mediates the response of Arabidopsis cells to the inhibition of cellulose synthesis. Current Biology. 2007;**17**(11):922-931

[25] Barry CS, Giovannoni JJ. Ethylene and fruit ripening. Journal of Plant Growth Regulation. 2007;**26**: 143-159

[26] Petit J, Bres C, Reynoud N, Lahaye M, Marion D, Bakan B, et al. Unraveling cuticle formation, structure, and properties by using tomato genetic diversity. Frontiers in Plant Science. 2021;**12**:778131

[27] Huang H, Jiang Y. Chemical composition of the cuticle membrane of pitaya fruits (Hylocereus Polyrhizus). Agriculture. 2019;**9**(12):250

[28] Belge B, Llovera M, Comabella E, Graell J, Lara I. Fruit cuticle composition of a melting and a nonmelting peach cultivar. Journal of Agricultural and Food Chemistry. 2014;**62**(15):3488-3495

[29] Huang H, Burghardt M, Schuster A-C, Leide J, Lara I, Riederer M. Chemical composition and water permeability of fruit and leaf cuticles of Olea europaea L. Journal of Agricultural and Food Chemistry. 2017;**65**(40):8790-8797

[30] Tafolla-Arellano JC, Zheng Y, Sun H, Jiao C, Ruiz-May E, Hernández-Oñate MA, et al. Transcriptome analysis of mango (Mangifera indica L.) fruit epidermal peel to identify putative

cuticle-associated genes. Scientific Reports. 2017;7(1):46163

[31] Matas AJ, Yeats TH, Buda GJ, Zheng Y, Chatterjee S, Tohge T, et al. Tissue-and cell-type specific transcriptome profiling of expanding tomato fruit provides insights into metabolic and regulatory specialization and cuticle formation. The Plant Cell. 2011;**23**(11):3893-3910

[32] Alkio M, Jonas U, Sprink T, van Nocker S, Knoche M. Identification of putative candidate genes involved in cuticle formation in Prunus avium (sweet cherry) fruit. Annals of Botany. 2012;**110**(1):101-112

[33] Martínez-López LA, Ochoa-Alejo N, Martínez O. Dynamics of the chili pepper transcriptome during fruit development. BMC Genomics. 2014;**15**(1):1-18

[34] Wang Y, Li W, Chang H, Zhou J, Luo Y, Zhang K, et al. SRNAome and transcriptome analysis provide insight into strawberry fruit ripening. Genomics. 2020;**112**(3):2369-2378

[35] Vergara-Pulgar C, Rothkegel K, González-Agüero M, Pedreschi R, Campos-Vargas R, Defilippi BG, et al. De novo assembly of Persea americana cv.'Hass' transcriptome during fruit development. BMC Genomics. 2019;**20**:1-14

[36] Li C, Wang Y, Huang X, Li J, Wang H, Li J. De novo assembly and characterization of fruit transcriptome in Litchi chinensis Sonn and analysis of differentially regulated genes in fruit in response to shading. BMC Genomics. 2013;**14**:1-16

[37] Saladié M, Matas AJ, Isaacson T, Jenks MA, Goodwin SM, Niklas KJ, et al. A reevaluation of the key factors that influence tomato fruit softening and integrity. Plant Physiology. 2007;**144**(2):1012-1028

[38] Lara I, Heredia A, Domínguez E. Shelf life potential and the fruit cuticle: The unexpected player. Frontiers in Plant Science. 2019;**10**:770

[39] Romero P, Rose JK. A relationship between tomato fruit softening, cuticle properties and water availability. Food Chemistry. 2019;**295**:300-310

[40] Suh MC, Samuels AL, Jetter R, Kunst L, Pollard M, Ohlrogge J, et al. Cuticular lipid composition, surface structure, and gene expression in Arabidopsis stem epidermis. Plant Physiology. 2005;**139**(4):1649-1665

[41] Quinet M, Angosto T, Yuste-Lisbona FJ, Blanchard-Gros R, Bigot S, Martinez J-P, et al. Tomato fruit development and metabolism. Frontiers in Plant Science. 2019;**10**:1554

[42] Bajar A, Podila GK, Kolattukudy P. Identification of a fungal cutinase promoter that is inducible by a plant signal via a phosphorylated trans-acting factor. National Academy of Sciences of the United States of America. 1991;**88**(18):8208-8212

[43] Kolattukudy P. Lipid-derived defensive polymers and waxes and their role in plant–microbe interaction. In: Lipids: Structure and Function. Massachusetts, USA: Elsevier Inc.; 1987. pp. 291-314

[44] Woloshuk CP, Kolattukudy P. Mechanism by which contact with plant cuticle triggers cutinase gene expression in the spores of Fusarium solani f. sp. pisi. National Academy of Sciences of the United States of America. 1986;**83**(6):1704-1708

[45] España L, Heredia-Guerrero JA, Segado P, Benítez JJ, Heredia A, Domínguez E. Biomechanical properties

of the tomato (Solanum lycopersicum) fruit cuticle during development are modulated by changes in the relative amounts of its components. New Phytologist. 2014;**202**(3):790-802

[46] Johnson PR, Ecker JR. The ethylene gas signal transduction pathway: A molecular perspective. Annual Review of Genetics. 1998;**32**(1):227-254

[47] Arya GC, Sarkar S, Manasherova E, Aharoni A, Cohen H. The plant cuticle: An ancient guardian barrier set against long-standing rivals. Frontiers in Plant Science. 2021;**12**:663165

[48] Serrano M, Coluccia F, Torres M, L'Haridon F, Métraux J-P. The cuticle and plant defense to pathogens. Frontiers in Plant Science. 2014;**5**:274

Chapter 3

Impact of Ripening and Processing on Color, Proximate and Mineral Properties of Improved Plantain (*Musa spp AAB*) Cultivars

Ekpereka Oluchukwu Anajekwu, Alamu Emmanuel Oladeji, Wasiu Awoyale, Delphine Amah, Rahman Akinoso and Busie Maziya-Dixon

Abstract

Recently breeders have developed high-yielding and disease-resistant hybrid plantain varieties that need evaluation for end-use. This study evaluated the effect of ripening and processing methods on the color, nutritional, and mineral properties of hybrid plantain cultivars. Plantain pulps were subjected to frying (170°C for 2 min), boiling (100°C for 15 min), and drying (65°C for 48 h) at unripe and ripe stages before analysis. Ripening and processing methods had a highly significant ($p < 0.05$) effect on all color and nutritional composition but significant on some minerals such as potassium. There was an interactive effect between ripening and processing methods on all color parameters and nutritional composition. In conclusion, fried samples recorded the highest mineral composition and vitamin C values, while boiled samples had the highest total carotenoid and color properties. Unripe plantain showed the most increased potassium, magnesium, calcium, and sodium contents. Processing plantain fruits at unripe stages were the best option to optimize nutrient availability.

Keywords: plantain, ripeness, processing, color, nutritional qualities

1. Introduction

Plantain (*Musa parasidiaca*) is a staple food consumed throughout the tropics aside from maize, rice, and wheat and is consumed as an essential food crop in West and Central Africa [1, 2]. The world production of plantain is over 76 million metric tons, out of which over 12 million metric tons are harvested in Africa yearly. Despite its prominence, Nigeria does not feature among plantain exporting nations because it has more for local consumption than export. Nigeria is one of the largest plantains producing countries in the world and have been regarded as a primarily starchy staple from some food consumption surveys [3]. It has been reported that in Nigeria, plantain fruits can be consumed unripe (green), yellowish-green (fairly ripe), or fully ripe after boiling, frying and drying/milling into flour to be taken with vegetable sauce after turning the flour in boiling water to form a textured dough [4]. Households in Nigeria consume different varieties of plantain. Still, the most preferred (plantain) varieties are the type locally known as "*Agbagba*" (*Musa paradisiaca*) is a significant food and is particularly desired for the variability in the stages of ripeness and cooking methods [5].

Processing plantains before storage rather than directly storing freshly harvested fruit has been advocated as one of the alternatives to reduce postharvest losses. Processing improves the digestibility of foods, promotes palatability, improves their keeping quality, and makes the food safer. However, processing may affect food's nutritional and mineral composition. Many nutritive minerals are essential to living organisms because they activate enzymes, hormones, and other organic molecules that participate in the growth, function, and maintenance of life processes. The consumption of plantain have been reported to provide an invaluable source of carbohydrate and caters to the calorific need of many developing countries [6]. According to Anajekwu *et al.* [2], plantain are a good source of mineral especially potassium in the diet but low in protein, fat and vitamin C. Ayodele *et al.* [7] reported a significant variation in the proximate composition of plantain fruits following the natural ripening and cooking (processing) method employed.

Plantains contains high amount of bioactive compounds like phenolic compounds, flavonoids, carotenoids and bio-genic amines. Phenolic compounds such as gallic acid, catechin, epicatechin, tannins and anthocyanins. The bioactive compounds have a clear therapeutic benefits to human health by contributing towards antioxidant activities [8]. These bioactive compounds are present in the raw and ripened plantains. Plantain pulps also contain a low levels of some phytosterols. Carotenoids are pro-vitamins which provide health benefits due to their physiological functions and serve as an antioxidants that scavenge singlet oxygen released in human body. Phytosterols are naturally plant sterols which serve as a functional ingredient that have health benefits such as lowering blood cholesterol level and its absorption in the intestine. Generally, the antioxidant activities of these bioactive compounds play a major role in reducing the risk of diseases such as, diabetes, heart problems and eye diseases which is on the rise throughout the world [9].

Black Sigatoka disease *Mycosphaerella fijiensis* (a *Musa* leaf spot disease) in the early 1980s threatened the livelihood and welfare of millions of sub-Saharan Africa, of which Nigeria is inclusive [2]. International Institute of Tropical Agriculture (IITA), in collaboration with other International Agricultural Research Centers (IARCs), developed varieties of hybrid cultivars of plantain and banana that are pest and disease resistant and high yielding, combined with good postharvest qualities to counteract the severe threat posed by Black Sigatoka diseases to *Musa spp* production in Nigeria [2].

However, until now, published studies are scarce on the effect of ripening and processing methods on the nutritional, mineral, color, dry matter, ascorbic acid (Vitamin C), and total carotenoid content of hybrid plantain cultivars. Therefore, the main objective of this study is to determine the impact of ripening and processing methods on the color, nutritional and mineral composition of selected hybrid plantain cultivars.

2. Materials and methods

2.1 Materials

The selected hybrid plantain cultivars were harvested from the International Institute of Tropical Agriculture (IITA) plantain and banana research farm, Ibadan, Oyo state, South/West Nigeria (7°22′N, 3°58′E, altitude 225 m). Four plantain varieties were evaluated: *PITA 26* and *PITA 27*, *Mbi egome*, and *Agbagba* (control). The selected plantain cultivars were divided into two batches; a batch was processed at the matured green (stage 1) and the other at ripe yellow (allowed to ripen after harvesting to stage 5). The ripening stages of the fruits were determined following a ripening chart [10]. The chemicals used for the research study were of analytical grade. All analyses were carried out in duplicates at the Food and Nutrition Sciences Laboratory of IITA, Ibadan, Oyo state.

2.2 Frying of plantain fingers

Unripe and ripe plantain bunches were separated into individual fingers, washed, peeled, and then sliced longitudinally into small round slices (2 mm thickness) with the aid of a sterile stainless-steel kitchen knife and fried in vegetable oil (specific gravity of 0.92 g/cm^3) for 2 minutes at 170°C. After frying, it was cooled and packaged in a polyethylene bag for further analysis at 30 ± 2°C [11].

2.3 Boiling of plantain fingers

Fingers of plantain were selected randomly from each set (unripe and ripe) and cut into a cooking pot containing 2 L of water each and then cooked for 15 min at 100°C. The peel of the plantain was not removed to prevent the leaching of nutrients into the boiling water during cooking. After the boiling, each set of plantain was drained out of hot water and cooled for 10 min before the peels were removed according to the method described by [12] with slight modification.

2.4 Production of plantain flour by drying

The method described by Anajekwu *et al.* [2] was used to prepare the Plantain flour. The plantain fruits were removed from bunches, washed, peeled manually, sliced (2 mm thickness) using a stainless-steel kitchen slicer, blanched at 80°C for 5 min, and dried in a cabinet drier at 65°C for 48 h. The dried sliced plantain fruits were milled, sieved, packaged in a low-density polyethylene bag, sealed, and stored for subsequent use.

2.5 Chemical composition of plantain

The plantain samples' moisture, ash, fat, protein, fiber, sugar, starch, and dry matter contents were determined as described by AOAC methods.

2.5.1 Moisture and dry matter content

Moisture content was determined using the method of AOAC and Idowu [13, 14]. Plantain sample (5 g) was weighed into a pre-weighed clean dried dish, after which the dish was placed in a well-ventilated oven (UF55 Memmert Oven model) maintained at 103 ± 2°C for 16 h. Transferred to the desiccator at room temperature to cool. After cooling for about 30 min, it was weighed as quickly as possible. The loss in weight was recorded as moisture and moisture content was calculated using the equation below:

$$\text{Moisture content} = \frac{M_1 - M_2 \times 100}{M_1 - M_0} \quad (1)$$

where M_0 = Weight in g of dish.
M_1 = Weight in g of dish and sample before drying.
M_2 = Weight in g of dish and sample after drying.
Note that M_1-M_0 = weight of sample prepared for drying.
% Dry Matter Content = 100 - % Moisture Content.

2.5.2 Ash content

This was determined by the method of AOAC and Idowu [13, 14]. It involves burning off all organic constituents at 600°C for 6 h in a furnace (VULCANTM furnace model 3–1750). Crucibles were washed, dried and allowed to cool in the desiccators. Each sample (3 g) was weighed into weighed crucibles. The weight of the residue after incineration was recorded as the ash content.

$$\text{Ash content} = \frac{W_3 - W_1 \times 100}{W_2} \quad (2)$$

W_3 = Weight of crucible+ ash.
W_2 = Weight of sample only.
W_1 = Weight of empty crucible.

PITA 27(Hybrid)   PITA 26 (Hybrid)   Mbi Egome   Agbagba

2.5.3 Protein content

Crude protein was determined using Kjeldahl method ([15], Method 46-12.01) [14]. Exactly 0.2 g of sample was weighed in a paper and transferred into a digestion tube and one tablet of Kjeldahl catalyst (copper) and 4 ml of conc. H_2SO_4 were added. This was transferred into a fume cupboard and 4 ml of H_2O_2 was added, fuming was allowed to stop. The mixture was placed on Tecator digestion block pre-set at 420°C and digested for 1 h; at the end of which all organically-bound nitrogen was converted to Ammonium Hydrogen Sulphate. With the addition of a strong alkali (NaOH, 40%) and the application of heat, ammonia NH_3 was distilled out, and collected in 1% boric acid receiver solution containing Bromocresol green/methyl red mix indicator. Blanks (Paper plus Kjeldahl catalyst) were prepared and treated similarly. Rack of digestion tubes was removed from the block and allowed to cool to room temperature.

The tube containing the blank sample was placed in the distillation unit of the system, and the weight of the sample to be analyzed was entered using the key board on the system and the system was programmed to automatically perform the distillation and titration of the sample. Likewise, in turns, the tubes containing the samples digest were placed in the distilling unit of the system. The system was also programmed to automatically perform the distillation and titration. Results were displayed automatically at the end of each analysis according to the following equation:

Calculation:

$$\%\text{Nitrogen} = \frac{(\text{sample titre} - \text{blank titre}) \times M \times 14.007 \times 100}{\text{Sample weight}} \qquad (3)$$

% Protein (crude) = % g Nitrogen x Conversion factor.
M = Molarity of the acid.

2.5.4 Fat content

An automated method (Soxtec™ 8000 System, [14, 15]) was used to determine crude fat. About 3 g of samples was weighed; transferred into a clean thimble plugged with cotton wool and inserted into the Soxtec™ 8000 apparatus. Clean pre-weighed extraction cup containing 50 ml n-hexane was placed on the heating mantle of the apparatus previously heated up to 120°C and then the thimble containing the sample was lowered into it. This set up was left in this boiling position for 15 minutes. After the extraction the thimble (i.e. sample) was lifted up and left in the rinsing position for 45 minutes. Thereafter, air knob was turned on and the hexane was allowed to evaporate for some 10 m minute. Extraction cup was further dried in hot-air oven for 20–30 minutes at 105°C to rid it of residual hexane. This was cooled in the desiccator and weighed. Fat content was calculated as follows:

$$\%\text{Fat} = \frac{W_3 - W_2 \times 100}{W_1} \qquad (4)$$

W_3 = Weight of the cup with extracted oil.
W_2 = Weight of the empty cup.
W_1 = Weight of sample.

2.5.5 Fiber content

An automated method (Fibertec™, [13]) was used to determine crude fiber content. About 0.5 g of celite and 1 g of already defatted samples was weighed into the pre-weighed glass crucibles. The equipment was turned on and R1 was pressed to heat the reagents. The crucibles were inserted using the holder and lock into position in front of the radiator in the Fibretec hot extraction unit ensuring that the safety latch engages. Placed the reflector in front of the crucibles and put all the valves in closed position. The cold water tap was opened for 1–2 minutes for reflux system. The preheated reagents were added into each column and 2–4 drops of n-octanol was added to prevent foaming, turned on the heater control fully clockwise. When the reagents started boiling, adjustment was made to moderate boiling using the heater control. The boiling time was measured from the time when the solution has reached the boiling point. At the end of extraction, the heater was turned off and placed the valves in Vacuum position and the cold water tap was opened to full flow rate for the water suction pump and started filtration. The crucible containing the residue was released with the safety hook. The crucible containing residue was dried overnight in 70°C vacuum oven. Cooled in desiccator and took the weight. The crucible containing residue was incinerated for at least 3 h at 525 ± 15°C, removed crucible from the furnace after it is cooled down 250°C. Cooled in desiccator and weighed.

$$\text{\%Crude fiber} = \text{Difference in weight} \times 100. \tag{5}$$

2.5.6 Carbohydrate content

$$\text{\%Carbohydrate} = 100 - (\text{Moisture} + \text{Ash} + \text{Fat} + \text{Protein} + \text{Fiber}) \text{ contents}. \tag{6}$$

2.5.7 Starch and sugar content

This was carried out according to the method described by AOAC and Idowu [13, 14]. Finely ground sample (0.02 g) was weighed into centrifuge tubes and 1 ml of 95% ethanol was added, followed by 2 ml of distilled water and 10 ml hot ethanol. The mixture was vortexed and centrifuged at 2000 rpm for 10 min. The supernatant was collected and used for free sugar analysis, while the residue was used for starch analysis.

To the residue 7.5 ml of concentrated perchloric acid was added and allowed to hydrolyze for 1 h. It was then diluted to 25 ml with distilled water and filtered through Whatman No. 2 filter papers. From the filtrate 0.05 ml was taken, made up to 1 ml with distilled water, vortexed and the color was developed by adding 0.5 ml phenol followed by 2.5 ml of conc. H_2SO_4. This was vortexed, allowed to cool to room temperature and the absorbance was read on a spectrophotometer (Genesys 10S UV–VIS, USA, Model) at 490 nm. To the supernatant made up to 20 ml with distilled water, an aliquot of 0.2 ml was taken, 0.5 ml (5%) phenol and 2.5 ml conc. H_2SO_4 was added. This was allowed to cool and the absorbance read at 490 nm.

The glucose standard solution was prepared by weighing 0.01 g of D-glucose into a 100 ml volumetric flask. This was dissolved and made up to 100 ml mark with distilled water. 0.1, 0.2, 0.3, 0.4 and 0.5 ml of the stock (100 μg/ml glucose) solution was dispensed into test tubes and each was made up to 1.0 ml with distilled water. This corresponds to 10, 20, 30, 40, and 50 μg glucose per ml. this was then followed by the addition of 0.5 ml of 5% phenol and 2.5 ml of H_2SO_4, vortexed, cooled and the absorbance read at 490 nm. Then a graph (standard glucose curve) of Absorbance against Concentration was plotted to determine the slope and intercept.

The sugar and starch content was determined by the calculation below:

$$\text{\%Sugar} = \frac{(A - I) \times D.F \times V \times 100}{B \times W \times 106} \tag{7}$$

$$\text{\%Starch} = \frac{(A - I) \times D.F \times V \times 0.9 \times 100}{B \times W \times 106} \tag{8}$$

A = Absorbance of sample
I = Intercept of sample
D.F = Dilution factor (depends on aliquot taken for assay)
V = Volume
B = Slope of the standard curve
W = Weight of the sample

2.6 Mineral composition of plantain

The mineral analysis was determined using an inductively coupled plasma optical emission spectrometer (ICP-OES) by the method described by AOAC [13] and Anajekwu et al. [2]. About 0.4–0.5 g of the sample was mixed with 2 ml of concentrated redistilled Nitric acid (HNO_3) in a 50 ml digestion tube and then left overnight for the cold digestion process. The mixture was placed in a digestion block at the temperature of 120°C starting and increased to 150°C. As the liquid dried off, 2 ml of concentrated Nitric acid was added until the solution was clear. A solution of 50/50 (v/v) Nitric acid and Perchloric acid was added, increasing the temperature to 180 to 220°C. The whitish ash-like residue was obtained, and the ash residue was dissolved using 1 ml of Hydrochloric acid (HCl) and 10 ml of 5% Nitric acid, vortexed, and transferred into 15 ml centrifuge tubes. The solution was injected into the ICP-OES to determine the mineral content. The mineral content (mg/kg) of each sample was calculated as follows:

$$\text{Mineral content (mg/kg)} = \frac{\text{Concentration (ppm) x DF}}{\text{Sample weight (g)}} \qquad (9)$$

D.F = Dilution factor

2.7 Color parameters

The method described by Anajekwu et al. [2] was used for the color parameters (L^*, a^*, and b^*) determination using a Chromater (Color-Tec-PCM TM, Omega Engineering Inc., Stanford, CT). The colorimeter was standardized, and samples were placed in the sample holder. The color measurement (L^*, a^*, b^*) was done in triplicates. The color intensity (ΔE) was calculated using the formular:

$$\Delta E = \left(\Delta L^2 + \Delta a^2 + \Delta b^2\right)^{1/2} \qquad (10)$$

2.8 Vitamin C content

Vitamin C content was determined according to the method described by Amoros et al. [16] with slight modification. 7.5 g of the sample was weighed and mixed directly in the extraction tube. 38 ml of extracting solution (0.4% oxalic acid and 20% acetone) was added and homogenized in the Ultra Turrax for 1 minute. The mixer with the extracting solution and the glass tube walls (approximately 3 ml) were rinsed. The extract was filtered through Whatman # 2 filter paper using a vacuum pump and rinsed the glass tube and filter walls. The filtered extract was transferred into a volumetric flask (protected from light) and brought to 50 ml with extracting solution and mixed. The reagent blank contains 1 ml of the extracting solution was mixed with 9 ml of 2,6 dichloroindophenol (DCIP) diluted solutions, and after 1 minute, the blank absorbance was read at 520 nm. About 1 ml of the extract (filtered solution of the sample) was with 9 ml of 2.6 DCIP diluted solutions, and the absorbance (Sample Absorbance) was read at 520 nm (read twice for each volumetric flask) after 1 minute. Sample blank contains 1 ml of the extract was mixed with 9 ml of distilled water, and the absorbance (Sample Blank: SB) was read at 520 nm. The sample blank was subtracted from the absorbance and then subtracted from the reagent blank. This final value (Real Absorbance) was used to estimate the concentration of ascorbic acid.

2.9 Total carotenoid composition

The method described by Anajekwu et al. [2] was used for total carotenoid determination. About 50 ml of cold acetone and 2 g of cellite were added to 5 g of plantain samples and crushed with a pestle. It was filtered with suction through the Buchner funnel using Filter paper (Whatman 90 mm). Washed mortar, pestle, and residue with small amounts of acetone, and the washings were received in the funnel. The crushing and filtration were repeated twice (until the residue was colorless). Petroleum ether was used for the extraction of carotenoids and quantified using a UV–VIS spectrophotometer at 450 nm.

Total carotenoid (TC spec) was calculated as follows:

$$TC(\mu g/g) = \frac{A\ total\ x\ volume\ (ml)\ x\ DF\ x\ 104}{A^{1\%}_{1cm}\ x\ sample\ weight\ (g)} \quad (11)$$

where, A_{total} = absorbance at 450 nm, DF = dilution factor, Volume (ml) = total volume of extract (25mls), $A^{1\%}_{1cm}$ = 2592 (absorption coefficient of beta-Carotene in petroleum ether (PE).

2.10 Statistical analysis

The data obtained in the laboratory were subjected to analysis of variance (ANOVA) using the Statistical Analytical System (SAS) package (SAS 9.3 version) [17], and the means were separated using Least Significant Difference (LSD). The significance test was done at the 5% probability level ($p < 0.05$).

3. Results and discussion

3.1 Color parameters of plantain cultivars as affected by ripening and processing methods

Table 1 shows the color parameters of PITA 26, PITA 27, Mbi Egome and Agbagba fruit pulp as affected by ripening and processing methods. The analysis of variance (ANOVA) result showed that varieties, processing methods, ripening stages and had a highly significant effect ($p < 0.001$) on the color properties (L^*, a^*, b^* and ΔE) of the plantain varieties studied. Color parameters significantly increased with ripeness ($P < 0.05$), particularly for PITA 26 and PITA 27. The lightness L^* ranged from 26.55–62.88 for PITA 26, 29.99–63.24 for PITA 27, 28.18–68.15 for Mbi Egome and 32.05–65.60 for Agbagba. Dried fruits had the highest lightness for all ripening stages, followed by boiled fruits. The drying process significantly impacted increased lightness, which could be attributed to the loss of moisture and non-enzymatic Maillard browning, which occurred under the conditions prevailing during the drying process, favoring color change [18]. Similar result of lightness (L > 50) was observed for plantain flour reported by Fadimu et al. [18]. Mba et al. [19] observed that at all the ripening stages, evaluated fried plantain chips had bright colors (L > 50) which is

Sample/Ripening stage	Processing Methods	L*	a*	b*	ΔE
PITA 26 Unripe (Stage 1)	Frying	30.18[b]	4.28[f]	11.07[c]	1.79[a]
	Drying	62.88[g]	−4.17[b]	12.12[d]	34.74[g]
	Boiling	49.78[d]	−4.91[b]	18.59[f]	23.34[e]
	Raw	49.43[d]	0.17[d]	16.68[e]	21.20[d]
PITA 26 Ripe (Stage 5)	Frying	26.55[a]	2.67[e]	8.34[a]	4.55[b]
	Drying	60.12[f]	−7.11[c]	10.22[b]	33.17[g]
	Boiling	55.60[e]	−7.51[c]	23.73[g]	31.05[f]
	Raw	44.99[c]	−0.15[a]	16.42[e]	16.98[c]
PITA 27 Unripe (Stage 1)	Frying	38.84[b]	0.17[e]	14.58[c]	10.68[b]
	Drying	63.24[h]	−5.22[c]	13.36[b]	35.36[g]
	Boiling	53.80[e]	−5.63[c]	21.37[e]	28.01[e]
	Raw	48.50[d]	−0.01[a]	15.72[d]	20.15[d]
PITA 27 Ripe (Stage 5)	Frying	29.99[a]	−0.49[b]	3.35[a]	9.65[a]
	Drying	61.32[g]	−7.34[d]	3.33[a]	39.42[h]
	Boiling	58.83[f]	−7.41[d]	22.86[f]	33.53[f]
	Raw	41.57[c]	−0.57[b]	14.76[c]	13.47[c]
Mbi Egome Unripe (Stage 1)	Frying	52.38[c]	−4.48[b]	26.95[g]	28.82[e]
	Drying	67.00[f]	−6.78[d]	17.26[c]	39.71[g]
	Boiling	53.17[d]	−2.98[a]	22.74[f]	27.20[d]
	Raw	50.35[b]	2.33[g]	20.05[d]	22.82[b]
Mbi Egome Ripe (Stage 5)	Frying	28.18[a]	2.04[f]	10.46[b]	2.31[a]
	Drying	68.15[g]	−5.16[c]	7.45[a]	38.14[f]
	Boiling	61.93[e]	−4.26[b]	31.60[h]	39.05[g]
	Raw	50.80[b]	1.75[e]	21.12[e]	23.67[c]
Agbagba Unripe (Stage 1)	Frying	52.94[d]	1.31[b]	26.59[g]	29.07[d]
	Drying	65.60[h]	−5.55[c]	13.69[c]	37.76[h]
	Boiling	59.74[f]	−3.73[a]	25.98[f]	34.49[f]
	Raw	48.02[b]	1.82[f]	17.97[d]	19.99[b]
Agbagba Ripe (Stage 5)	Frying	32.05[a]	1.13[d]	7.25[b]	6.20[a]
	Drying	63.11[g]	−3.20[a]	5.20[a]	36.48[g]
	Boiling	54.58[e]	−3.46[a]	27.25[h]	30.54[e]
	Raw	50.87[c]	1.45[e]	19.08[e]	23.06[c]

Means followed by different superscripts within a column indicate a significant difference ($p < 0.05$). L* = lightness, a* = Redness, b* = Yellowness, ΔE = Color intensity.

Table 1.
Mean value of color parameters of plantain cultivars by ripening and processing methods.

similar to results generated for Mbi egome and Agbagba fried unripe plantain chips in this study.

However, the lightness of the fried chips significantly decreased as ripening progressed [19], this is similar to the results in this study. Frying reduced the lightness of the fruits. Redness a* varied across the ripening stages and processing methods. Redness significantly ($P < 0.05$) decreased with ripeness for PITA 26 and PITA 27 but increased with ripeness for Mbi egome and Agbagba which is similar to the report by Mba et al. [19]. Fried fruits had the highest redness for all ripening stages. Yellowness b* ranged from 8.34–18.59 for PITA 26, 3.33–22.86 for PITA 27, 7.45–31.60 for Mbi Egome and 5.20–27.25 for Agbagba. Boiling increased the yellowness of fruits at all ripening stages. The more desired color of fried chips is the yellowness and the decrease in lightness positively correlated with increase in redness a* and yellowness b* of fried chips with time [19]. The data also showed that frying was significantly affected by ripening. The color intensity ΔE ranged from 1.79–34.74 for PITA 26, 9.65–39.42 for PITA 27, 2.31–39.71 for Mbi Egome and 6.20–37.76 for Agbagba. Boiled and dried fruits had the highest color intensity at all ripening stages. Fried fruits recorded the least color intensity.

3.2 Chemical composition of plantain cultivars as affected by ripening and processing methods

Table 2 shows the chemical compositions of selected plantain cultivars as affected by the ripening and processing method. The analysis of variance (ANOVA) showed that variety had a significant ($p < 0.001$) effect on all the chemical properties except protein. However, the processing method and ripening stage had a highly significant ($p < 0.001$) effect on the chemical components of the plantain cultivars. However, the second level interaction of variety x processing method, processing method x ripening stage, and variety x processing method had a significant ($p < 0.001$) on all the chemical parameters, except a slight significance ($p < 0.05$) for protein.

It implies that variety, processing method, and ripening stage affect the chemical contents of plantain cultivars. Among the explanatory variables, the variable variety is more influential for all the chemical components except protein, where the processing method and ripening stage are the most influential variables. Thus, the processing methods and ripening stages strongly affected the concentration of ash, fat, protein, carbohydrate, sugar, starch, moisture, vitamin C and total carotenoid content of plantain cultivars. The result agrees with what Ayodele et al. [7] reported that the natural ripening and cooking method significantly affect the proximate composition of plantain fruits. The mean separation results showed a significant difference ($P < 0.05$) for most of the proximate components of plantain cultivars studied. Moreover, the ash, protein, total carbohydrate, vitamin C, and starch contents decreased with ripeness, whereas fat, sugar, and total carotenoid contents increased notably at the fully ripe stage(stage 5).

Ash content reflects mineral status and the inorganic residue after the water and organic matter have been removed by burning a food sample [20]. Ash content ranged from 1.20–3.33 g/100 g for PITA 26, 1.29–3.69 g/100 g for PITA 27, 1.05–2.09 g/100 g for Mbi Egome and 1.06–3.34 g/100 g for Agbagba respectively. Dried fruits had the highest ash content for PITA 26, PITA 27 and Agbagba at the unripe stages. However, boiled fruits had the least ash content for Mbi Egome and Agbagba. The ash contents

Sample/Ripening stage	Processing Methods	Ash	Fat	Protein	CHO	Sugar	Starch	MC	Vit. C (mg/100 g)	TC µg/g	Fiber
PITA 26 Unripe (Stage 1)	Frying	2.29^d	17.91^f	1.43^d	68.91^e	8.82^e	91.54^h	8.60^c	8.29^e	1.19^a	0.88^d
	Drying	3.33^e	1.20^d	2.93^f	85.66^g	6.29^b	89.37^g	6.15^a	10.25^f	2.92^b	0.73^c
	Boiling	1.53^b	0.72^b	0.49^a	28.17^c	2.00^a	49.12^c	68.56^f	7.31^d	4.18^c	0.53^a
	Raw	1.25^a	1.69^f	0.90^c	27.84^c	2.10^a	46.90^c	67.13^h	14.94^g	1.71a	1.19^f
PITA 26 Ripe (Stage 5)	Frying	1.65^c	28.97^g	1.90^e	41.27^d	18.99^g	35.14^a	25.53^d	5.49^b	1.99^a	0.68^b
	Drying	1.51^b	1.33^d	0.65^b	82.21^f	11.66^f	83.73^f	7.24^b	6.21^c	2.09^b	0.70^c
	Boiling	1.26^a	0.79^c	0.52^a	23.42^a	6.97^c	56.98^e	72.84^h	4.23^a	5.04^d	1.17^e
	Raw	1.20^a	0.60^a	0.70^b	26.82^b	7.58^d	37.28^b	69.38^g	7.71^d	4.92^c	1.30^g
PITA 27 Unripe (Stage 1)	Frying	2.48^d	11.31^c	1.55^e	74.43^f	5.39^b	80.07^f	9.48^b	9.13^d	5.89^d	0.75^b
	Drying	3.69^e	0.78^a	2.99^f	85.27^h	8.59^c	81.22^g	6.47a	10.97^e	5.18^d	0.81^c
	Boiling	1.64^b	0.47^a	0.69^b	28.50^c	1.78^a	57.24^d	67.42^g	7.43^b	6.85^e	1.28^d
	Raw	1.49^b	0.99^a	0.90^c	29.68^d	1.61^a	66.46^e	66.23^d	16.96^f	4.67^c	0.71^a
PITA 27 Ripe (Stage 5)	Frying	1.90^c	25.44^d	1.60^e	39.97^e	33.13^g	35.21^c	28.64^e	7.14^b	1.98^a	2.46^f
	Drying	1.66^b	2.02^b	1.25^d	80.15^g	13.79^e	84.45^h	6.24^a	8.18^c	3.21^b	0.66^a
	Boiling	1.41^b	0.81^a	0.58^a	19.92^a	12.88^d	30.43^b	75.14^h	6.06^a	5.28^d	2.50^g
	Raw	1.29^a	0.72^a	0.54^a	27.46^b	14.65^f	29.96^a	68.61^a	9.40^d	7.12^f	1.39^e
Mbi Egome Unripe (Stage 1)	Frying	1.60^b	11.94^e	2.02^d	78.56^g	6.30^b	85.47^g	5.20^a	7.89^b	5.43^e	0.68^c
	Drying	2.01^c	0.80^c	2.86^e	86.95^h	9.33^c	84.34^f	6.48^b	8.27^c	18.08^f	0.90^e
	Boiling	1.05^a	0.62^b	0.53^a	30.58^b	1.90^a	43.62^c	64.91^f	6.02^a	24.19^h	2.31^g
	Raw	1.07^a	0.60^b	0.81^b	36.58^d	1.59^a	54.02^e	60.31^d	9.42^d	19.77^g	0.64^b
Mbi Egome Ripe (Stage 5)	Frying	1.44^b	23.68^f	1.90^c	64.56^e	23.21^f	51.02^d	7.17c	7.35^b	6.29^b	1.25^f
	Drying	1.66^b	2.50^d	1.77^c	75.60^f	16.82^e	85.33^g	6.42^b	8.30^c	10.42^e	0.85^e
	Boiling	2.09^c	0.12^a	0.65^b	23.21^a	6.56^b	33.23^a	73.35^g	7.31^b	13.58^d	0.58^a
	Raw	1.06^a	0.50^b	0.86^b	34.63^c	10.28^d	39.54^b	62.16^e	10.37^c	14.72^c	0.78^d
Agbagba Unripe (Stage 1)	Frying	2.73^d	11.27^e	1.76^e	73.94^f	8.95^c	90.13^g	9.29^c	9.21^b	5.98^b	1.01^c
	Drying	3.34^e	0.49^b	2.47^e	85.46^h	6.75^a	93.96^h	7.27^a	10.54^c	4.87^b	0.97^c
	Boiling	1.06^a	1.82^c	0.76^b	33.07^c	6.34^a	67.33^d	62.78^f	9.04^b	17.42^e	0.52^a
	Raw	1.50^b	0.15^a	1.00^b	37.52^d	6.30^a	78.80^e	56.69^d	14.19^d	11.46^d	1.08^c
Agbagba Ripe (Stage 5)	Frying	2.69^c	22.18^f	1.48^c	39.39^e	20.13^f	61.48^c	32.58^c	8.70^a	7.06^c	1.69^d
	Drying	2.01^c	2.11^d	1.37^c	81.06^g	13.10^e	82.11^f	7.12^a	8.96^a	11.23^d	0.48^a
	Boiling	1.21^b	0.24^a	0.82^a	26.64^a	9.26^c	32.88^a	69.23^g	8.69^a	19.51^g	1.87^e
	Raw	1.42^b	0.13^a	0.36^a	34.81^b	10.50^d	35.34^b	62.58^e	9.50^b	18.81^f	0.70^b

Means followed by different superscripts within a column indicate a significant difference ($p < 0.05$). TC = Total Carotenoid, MC = Moisture content, Vit = Vitamin, CHO = Carbohydrate. The fiber content ranged from 0.53–1.30 g/100 g for PITA 26, 0.66–2.50 g/100 g for PITA 27, 0.58–2.31 g/100 g for Mbi Egome and 0.52–1.87 g/100 g for Agbagba respectively.

Table 2.
Mean value of some chemical composition of plantain cultivars (g/100 g) by ripening and processing methods.

of the raw plantain fruits at both unripe and ripe stages is similar to the result reported by Okareh et al. [21].

Fat contents ranged from 0.60–28.97 g/100 g for PITA 26, 0.47–25.44 g/100 g for PITA 27, 0.12–23.68 g/100 g for Mbi Egome, 0.13–22.18 g/100 g for Agbagba respectively and significantly ($P < 0.05$) increased with ripeness. Fried fruits had the highest content of fat at the ripe and unripe stages for all cultivars, which can be attributed to the addition of fat from the frying oil. A similar higher result was reported by Omotosho et al. [22] for plantain chips stating that frying influenced the fat level of plantain with vegetable oil. Agbagba cultivar (control) had the least fat content at the raw unripe and ripe stages compared to the other cultivars. Boiled and dried fruits had relatively low fat across the ripening stages and processing methods.

Protein content was not significantly influenced by ripeness and the processing methods employed. Still, it was correspondingly low in all the boiled fruits across the ripening stages for all the cultivars studied. The low protein content is characteristic of plantain [1, 20]. Protein content ranged from 0.49–2.93 g/100 g for PITA 26, 0.54–2.99 g/100 g for PITA 27, 0.53–2.86 g/100 g for Mbi Egome and 0.36–2.47 g/100 g for Agbagba respectively. Protein is essential in the human diet to grow and repair worn-out tissues. Dried fruits at the unripe stage had the highest protein for all cultivars studied, whereas fried fruits at the ripe stage recorded the least protein for all cultivars studied. Relatively high protein content in dried fruits may be due to protein modification or loss of moisture during the drying process, which eventually increases other nutrient compositions [20].

The carbohydrate levels were significantly ($P < 0.05$) higher at the unripe stages. Carbohydrate was relatively high in dried and fried fruits at all the fruit ripening stages but correspondingly low in boiled fruits. The carbohydrate content ranged from 23.42–85.66 g/100 g for PITA 26, 19.92–85.27 g/100 g for PITA 27, 23.21–86.95 g/100 g for Mbi Egome and 26.64–85.46 g/100 g for Agbagba respectively. Higher carbohydrates in dried and fried fruits may be due to moisture loss during drying and frying, which increases the concentration of soluble matter [20]. At the same time, the boiling process adds more moisture to the fruits leading to a drop in the concentration of soluble matters. Processing methods had been reported to improve carbohydrate availability in a more digestible form [23] which explained the significant increase in carbohydrates of dried and fried fruits. Similar result of carbohydrate for dried plantain flour was obtained as that reported by Okole et al. [20] and Fadimu et al. [18].

The moisture content of processed food gives an indication of its anticipated shelf life. Food with low moisture contents remains in good condition for a longer time than the one with high moisture content [20]. The moisture content significantly ($P < 0.05$) increased with ripening and was correspondingly low in dried and fried fruits across the ripening stages because of heat application. Moisture content was significantly ($P < 0.05$) high in boiled and raw fruits at all ripening) stages. The moisture ranged from 6.15–72.84 g/100 g for PITA 26, 6.24–75.14 g/100 g for PITA 27, 6.42–73.35 g/100 g for Mbi Egome and 7.21–69.28 g/100 g for Agbagba respectively. Okole et al. [20] reported high values of moisture content for dried plantain flour compared to that obtained in this study. A well dried food withstands microbial infestation better during storage.

The sugar content ranged from 2.00–18.99 g/100 g for PITA 26, 1.61–33.13 g/100 g for PITA 27, 1.59–23.21 g/100 g for Mbi Egome, 6.30–20.13 g/100 g for Agbagba and significantly ($P < 0.05$) increased with ripeness. Fried fruits at the ripe stage had the highest sugar content. Sugar was relatively high in fried and dried fruits at ripening stages but correspondingly low in boiled fruits. Raw fruits at the ripe stage showed higher sugar content than those at the unripe stage for all cultivars studied. As ripening continues, the starch content is broken down to sugar. As the ripening process increased, the starch content decreased. Fried and dried fruits had the highest starch at the unripe stage. The starch content ranged from 35.14–91.54 g/100 g for PITA 26, 29.96–81.22 g/100 g for PITA 27, 33.23–85.47 g/100 g for Mbi Egome and 32.88–93.96 g/100 g for Agbagba respectively. Raw fruits at the unripe stage of ripeness recorded higher starch content than the raw ones at ripe stages. As ripening continues, the starch content breaks down into sugars.

Vitamin C is a very good antioxidant, and highly cherished in mopping up reactive oxygen species in the body. Since the meals from this food sample are usually eaten with vegetables, it means this difference can always be made up from other ingredients used in cooking plantain meals [24]. Vitamin C content ranged from 4.23–14.94 mg/100 g for PITA 26, 6.06–16.96 mg/100 g for PITA 27, 6.02–10.37 mg/100 g for Mbi Egome and 8.69–14.19 mg/100 g for Agbagba respectively. Raw fruits at the unripe stage recorded the highest vitamin C for all cultivars except Mbi Egome. This may be due to varietal differences between the cultivars. Vitamin C was relatively high in dried and fried fruits at all the fruit ripening stages but correspondingly low in boiled fruits. This suggests that drying and frying significantly affected fruits by increasing the ascorbic acid because of moisture removal, which raised other nutritional components. Yarkwan and Uvir [24] reported vitamin C content of 5.00 mg/100 g for fresh unripe plantain which is lower than the values reported for raw unripe plantain in this study and this may be due to differences in variety of plantains.

Total carotenoid content significantly ($P < 0.05$) improved with ripening. Carotenoid is one of the most important classes of plant pigment and plays a critical role in defining the quality parameters of fruit and vegetables. The total carotenoid content ranged from 1.19–5.04 µg/g for PITA 26, 1.98–5.28 µg/g for PITA 27, 6.29–24.19 µg/g for Mbi Egome and 7.06–19.51 µg/g for Agbagba respectively. The fiber content ranged from 0.53–1.30 g/100 g for PITA 26, 0.66–2.50 g/100 g for PITA 27, O.58–2.31 g/100 g for Mbi Egome and 0.52–1.87 g/100 g for Agbagba respectively. Boiling seemingly increased the total carotenoid content across the fruit ripening stages. In most cases, boiled and dried fruits had a relatively higher total carotenoid content. The raw fruits at the ripe stage recorded higher total carotenoid content than the raw fruits at the unripe stage. The increase in carotenoid content observed in this study could be explained by the fact that maturation or ripening of fruits and vegetables is usually accompanied by enhanced carotenogenesis and provided the fruit or vegetable remains intact; this continues to occur even after harvest [25].

3.3 Mineral composition of plantain cultivars as affected by ripening and processing methods

Table 3 shows the effects of varying fruit ripening stages and processing methods on the mineral composition of PITA 26, PITA 27, Mbi Egome and Agbagba fruit pulp. Calcium content ranged from 7.82–10.85 mg/Kg for PITA 26, 8.00–11.98 mg/Kg for PITA 27, 7.43–10.30 mg/Kg for Mbi Egome and 7.48–10.41 mg/Kg respectively. Dried fruits had the highest concentration of calcium in all ripening stages. Fried fruits, in most cases, had a relatively high calcium content. Processing by boiling seemingly reduced the fruit's calcium content. Excessive level of calcium can cause constipation. This high calcium concentration might also impede the body's ability to absorb other minerals like iron and zinc [26]. Fruit Magnesium content significantly ($P < 0.05$)

Sample/Ripening stage	Processing Methods	Ca	Mg	K	Na	Mn	Fe	Cu	Zn	Al
PITA 26 Unripe (Stage 1)	Frying	10.74[d]	25.32[g]	365.20[g]	6.39[d]	0.16[b]	0.59[b]	0.19[a]	0.05[a]	0.86[d]
	Drying	10.77[d]	24.73[f]	366.45[g]	6.30[d]	0.15[b]	0.66[a]	0.25[b]	0.12[b]	0.59[b]
	Boiling	8.08[b]	8.56[a]	88.23[a]	5.29[b]	0.03[a]	0.37[a]	0.12[a]	0.01[a]	0.47[a]
	Raw	7.82[a]	9.73[b]	114.34[b]	5.01[a]	0.05[a]	1.53[d]	0.29[b]	0.01[a]	0.54[b]
PITA 26 Ripe (Stage 5)	Frying	10.25[d]	21.39e	311.60[f]	6.06[c]	0.13[b]	0.57[b]	0.23[b]	0.05[a]	0.66[c]
	Drying	10.85[d]	21.13[e]	309.23[c]	6.00[c]	0.11[b]	0.58[b]	0.31[c]	0.10[b]	0.61[b]
	Boiling	8.74[c]	12.45[c]	153.65[c]	5.53[b]	0.06[a]	0.56[b]	0.41[d]	0.02a	0.58[b]
	Raw	8.48[c]	13.51[d]	184.45[d]	5.45[b]	0.06[a]	0.72[c]	0.26[b]	0.03[a]	0.58[b]
PITA 27 Unripe (Stage 1)	Frying	10.39[e]	21.52[e]	358.85[g]	6.30[b]	0.09[b]	0.64[b]	0.16[a]	0.45[d]	0.62[a]
	Drying	11.98[f]	29.11[g]	528.50[h]	6.30[b]	0.03[a]	0.67[b]	0.40[c]	0.29[c]	0.68[b]
	Boiling	9.45[d]	9.78[a]	126.66[b]	5.80[a]	0.02[a]	0.47[a]	0.18[a]	0.24[c]	0.62[b]
	Raw	9.99[d]	12.26[d]	176.90[d]	5.60[a]	0.03[a]	0.49[a]	0.26[b]	0.01[a]	0.79[b]
PITA 27 (Ripe Stage 5)	Frying	9.13[c]	21.83[e]	346.95[f]	6.22[b]	0.07[b]	0.43[a]	0.23[b]	0.10[b]	0.50[a]
	Drying	10.12[e]	22.14[f]	351.10[e]	6.24[b]	0.04[a]	0.42[a]	0.33[c]	0.06[a]	0.41[a]
	Boiling	8.86[b]	10.62[b]	123.9[a]	5.82[a]	0.03[a]	0.92[c]	0.39[c]	0.01[a]	0.68[b]
	Raw	8.00[a]	11.62[c]	150.70[c]	5.48[a]	0.04a	0.55[b]	0.29[b]	0.02[a]	0.56[a]
Mbi Egome Unripe (Stage 1)	Frying	10.30[d]	23.67[c]	291.20f	6.20[c]	0.08[b]	0.77[b]	0.31[b]	0.08[b]	0.77[d]
	Drying	10.08[d]	28.99[f]	380.65[g]	6.48[c]	0.09[b]	0.85[b]	0.39[c]	0.18[b]	0.77[d]
	Boiling	9.51[c]	9.48[a]	79.30[a]	5.11[b]	0.02[a]	0.55[a]	0.26[a]	0.01[a]	0.61[c]
	Raw	8.85[b]	10.78[b]	100.59[d]	5.76[b]	0.02[a]	0.41[a]	0.22[a]	0.01[a]	0.52[b]
Mbi Egome Ripe (Stage 5)	Frying	8.63[b]	24.05[d]	287.65[e]	6.01[c]	0.08[b]	0.75[b]	0.27[a]	0.11[b]	0.53[b]
	Drying	8.61[b]	25.12[e]	290.22[f]	6.10[c]	0.10[b]	0.80[c]	0.31[b]	0.17[c]	0.56[b]
	Boiling	7.43[a]	9.91[a]	90.01[b]	5.20[b]	0.02[a]	0.49[a]	0.22[a]	0.03[a]	0.48[a]
	Raw	7.93[a]	10.06[b]	94.37[c]	4.72[a]	0.02[a]	0.48[a]	0.50[c]	0.07[a]	0.53[b]
Agbagba Unripe (Stage 1)	Frying	7.56[a]	23.83[f]	342.85[h]	6.66[b]	0.10[b]	0.58[a]	0.29[b]	0.12[b]	0.52[a]
	Drying	9.46[b]	27.25[g]	328.30[g]	7.24[c]	0.15[b]	0.84[c]	0.30[b]	0.19[c]	0.53[a]
	Boiling	7.58[a]	10.45[a]	119.27[b]	5.40[a]	0.02[a]	0.79[b]	0.13[a]	0.05[a]	0.64[b]
	Raw	7.93[a]	14.37[d]	160.10[d]	6.15[b]	0.04[a]	0.63[a]	0.27[b]	0.13[b]	0.61[b]
Agbagba Ripe (Stage 5)	Frying	10.30[c]	21.88[e]	268.65[f]	6.36[b]	0.14[b]	0.90[d]	0.45[c]	0.08[a]	0.76[c]
	Drying	10.41[c]	23.61[f]	264.11[e]	6.40[b]	0.16[b]	0.96[d]	0.51[c]	0.10[b]	0.77[c]
	Boiling	7.69[a]	11.25[b]	108.81[b]	5.54[a]	0.04[a]	0.53[a]	0.18[a]	0.02[a]	0.54[a]
	Raw	7.48[a]	12.36[c]	140.55[c]	5.36[a]	0.06[a]	0.45[a]	0.18[a]	0.04[a]	0.59[a]

Means followed by different superscript within a column indicate a significant difference ($p < 0.05$). Ca = calcium, Mg = magnesium, K=Potassium, Na = sodium, Mn = manganese, Fe = iron, Cu = copper, Zn = zinc, Al = Aluminum.

Table 3.
Mean value of mineral composition of plantain cultivars (mg/Kg) by ripening and processing methods.

decreased with ripeness, particularly for PITA 27, Mbi Egome and Agbagba cultivars. Boiling seemingly reduced the fruit's Magnesium content. Fried and dried fruits had the highest magnesium concentration in most ripening stages for all cultivars studied. Magnesium content ranged from 8.56–25.32 mg/Kg for PITA 26, 9.78–29.11 mg/Kg for PITA 27, 9.48–28.99 mg/Kg for Mbi Egome and 11.25–27.25 mg/Kg for Agbagba respectively.

The potassium content of the fruits was significantly influenced by fruit ripeness and the processing method. Still, there was a progressive decrease in potassium levels as ripening progressed across all processing methods studied. Potassium content ranged from 88.23–366.45 mg/Kg for PITA 26, 123.90–528.50 mg/Kg for PITA 27, 90.01–380.65 mg/Kg and 108.81–342.85 mg/Kg for Agbagba respectively. Inyang et al. [27] reported a similarly high value (476.09 mg/Kg) for dried plantain fruit. Dried fruits, followed by fried fruits, had the highest potassium content across the fruit ripening stages. High potassium is essential to help balance body sodium level to keep blood pressure from getting too high as well as playing an important role in heart, muscle and digestive function [26]. Potassium is required in a relatively large quantity in the human body because it functions as an electrolyte in the nervous system, helpful in osmoregulation and controlling high blood pressure.

Sodium ranged from 5.01–6.39 mg/Kg for PITA 26, 5.48–6.30 mg/Kg for PITA 27, 4.72–6.48 mg/Kg for Mbi Egome and 5.36–7.24 mg/Kg for Agbagba. Fried and dried fruits recorded the highest sodium at all ripening stages. Generally, the amount of sodium for all fruits at different processing methods and ripening stages were low compared to other nutrients like calcium, magnesium, and potassium. The low amount of sodium recorded in fruits makes them suitable for hypertensive patients [26].

Manganese, Iron, Zinc and Aluminum content was not significantly influenced by the stage of fruit ripeness and recorded low values across all the processing methods. Iron is a vital nutrient in the body but is relatively more minor in the samples studied. The highest Mn, Fe, Zn and Al concentrations were recorded for dried and fried fruits for all ripening stages. The boiling process significantly ($P < 0.05$) reduced the fruit Mn, Fe, Zn and Al contents, particularly at the ripe stages. Boiling had been implicated in reducing iron, copper, and zinc in plantain pulp.

The ANOVA results showed that variety had a no significant($P > 0.05$) effect on the minerals but a slightly significant ($p < 0.05$) effect on potassium However, the processing method had a significant ($p < 0.001$) effect on calcium, magnesium, potassium, and sodium but was not significant for other minerals. The effect of ripening on calcium, magnesium, potassium, manganese and aluminum contents was highly significant ($p < 0.001$) but had a slightly significant ($p < 0.01$) effect on iron and copper. Variety x processing method interaction had a slightly significant effect ($p < 0.05$) for potassium and manganese, but variety x ripening stage interaction had a slight significant effect ($p < 0.05$) on only potassium. However, it is imperative to note that the processing method x ripening stage interaction had no significant ($P > 0.05$) effect on all the minerals. The result agrees with banana varieties' reported results. Thus, the ripening stage is the only factor that showed a pronounced effect on all the minerals, while the processing method primarily affects the macro-elements (calcium, magnesium, potassium). This could be due to the leaching of those minerals into the water used for the processing, and these minerals have been identified as water-soluble.

Further studies should be conducted on the effect of processing methods on quality attributes and nutrient retention of other types of plantain hybrids and valorization of plantain peels used in this study. Other forms of processing methods not used in this study should be applied for further studies.

4. Conclusion

Generally, the plantain cultivars showed high moisture, carbohydrate and starch content but lower fat, ash, protein and sugar content. The unripe stage showed high ash, protein, carbohydrate and starch content but low moisture content, while the ripe stage showed high moisture, fat, and sugar content. At the unripe ripening stage, plantain showed the most increased potassium, magnesium, calcium, sodium, manganese, and iron mineral content over the ripe stage. Processing at the unripe stage is the best option to obtain the maximum nutrients in plantain fruits. Also, *Pita 26* (hybrid) showed the highest potassium, magnesium, calcium, sodium, manganese and iron mineral compositions over *Pita 27* (hybrid), *Mbi egome* and *Agbagba* (control) plantain cultivars. It implies that hybrid plantain can be commercialized because of the inherent nutrients after processing.

Moreover, the frying method maintained high ash, fat, carbohydrate, protein, sugar and starch content than other processing methods. Fried and dried fruits at the unripe stage kept high potassium, magnesium, calcium, manganese and low sodium mineral content over other processing methods; therefore, it stands as the best ripening stage.

Acknowledgements

The authors acknowledge the support received from the CGIAR Research Program on Roots, Tubers and Bananas (RTB), the staff of the Food and Nutrition Science Laboratory and the Plantain and Banana Breeding Unit of the International Institute of Tropical Agriculture (IITA), Ibadan, Nigeria, and Department of Food Technology of the University of Ibadan, Nigeria.

Funding details

This research received no specific grant from the public, commercial, or not-for-profit funding agencies.

Declaration of interest statement

The authors report there is no competing interest to declare.

CREDIT author statement

Oluchukwu Anajekwu was responsible for conceptualization, investigation, data curation, and writing original draft preparation. Emmanuel Alamu was responsible for resources, visualization, validation, writing the original manuscript, and writing, reviewing, and editing. Rahman Akinoso was accountable for supervising, methodology, investigation, writing, reviewing, and editing. Wasiu Awoyale was responsible for visualization, data curation, software, writing, reviewing, and editing. Delphine Amah was responsible for Resources, writing, reviewing and editing. Busie Maziya-Dixon was accountable for conceptualization, methodology, investigation, resources, supervision, writing, reviewing, and editing.

Limitation of the study

As a climacteric fruit, plantains (*Musa spp.*) undergo a fast-ripening process upon maturity by the release of ethylene gas leading to post harvest losses, and made it difficult to assess many hybrids and exploring other stages of ripening.

Author details

Ekpereka Oluchukwu Anajekwu[1,2], Alamu Emmanuel Oladeji[1,3]*, Wasiu Awoyale[1,4], Delphine Amah[1], Rahman Akinoso[2] and Busie Maziya-Dixon[1]*

1 International Institute of Tropical Agriculture (IITA), Oyo, Nigeria

2 Department of Food Technology, University of Ibadan, Oyo, Nigeria

3 International Institute of Tropical Agriculture, Southern Africa Research and Administration Hub (SARAH) Campus, Lusaka, Zambia

4 Department of Food Science and Technology, Kwara State University, Nigeria

*Address all correspondence to: oalamu@cgiar.org; b.maziya-dixon@cgiar.org

IntechOpen

© 2023 The Author(s). Licensee IntechOpen. This chapter is distributed under the terms of the Creative Commons Attribution License (http://creativecommons.org/licenses/by/3.0), which permits unrestricted use, distribution, and reproduction in any medium, provided the original work is properly cited.

References

[1] Adenitan A, Awoyale W, Akinwande AB, Maziya-Dixon B. Influence of drying methods on heavy metal composition and microbial load of plantain chips. Cogent Food and Agriculture. 2022;**8**:1-22

[2] Anajekwu EO, Maziya-Dixon B, Akinoso R, Awoyale W, Alamu EO. Physicochemical properties and total carotenoid content of high-quality unripe plantain flour from varieties of hybrid plantain cultivars. Journal of Chemistry. 2020;**2020**:1-7

[3] Ogechi UP, Akhakhia OI, Ugwunna UA. Nutritional status and energy intake of adolescents in Umuahia urban, Nigeria. Pakistan Journal of Nutrition. 2017;**6**(6):641-646

[4] Akinsanmi AO, Oboh G, Akinyemi JA, Adefegha AS. Assessment of the nutritional, anti-nutritional, and antioxidant capacity of unripe, ripe, and overripe plantain (*Musa paradisiaca*) peels. International Journal of Advanced Research. 2015;**3**(2):63-72

[5] Ubi GM, Nwagu KE, Jemide JO, Egu CJ, Onabe MB, Essien IS. Organoleptic and horticultural characterization of selected elite cultivars of plantain (*Musa paradisiaca L.*) for value addition and food security in Nigeria. Journal of Advances in Biology and Biotechnology. 2016;**6**(4):1-19

[6] Amah D, Stuart E, Mignouna D, Swennen R, Teeken B. End-user preferences for plantain food products in Nigeria and implications for genetic improvement. International Journal of Food Science and Technology. 2021;**56**(3):1148-1159

[7] Ayodele OD, Fagbenro I, Adeyeye A. The effect of processing method on the proximate, anti-nutrient and phytochemical composition of ripe and unripe plantain (*Musa paradisiaca*). Open Science Journal of Analytical Chemistry. 2019;**4**:1-6

[8] Singh JP, Kaur A, Shevkani K, Singh N. Influence of jambolan (*Syzgium cumini*) and xanthan gum incorporation on the physicochemical, antioxidant and sensory properties of gluten-free eggless rice muffins. International Journal of Food Science and Technology. 2015;**50**:1190-1197

[9] Balwinder S, Jatinder PS, Amritpal K, Narpinder S. Bioactive compounds in banana and their associated health benefits – A review. Food Chemistry. 2016;**206**:1-11

[10] Bhuiyan F, Campos NA, Swennen R, Carpentier S. Characterizing fruit ripening in plantain and Cavendish bananas: A proteomics approach. Journal of Proteomics. 2020;**214**:1-11

[11] Adeyanju JA, Olajide JO, Adedeji AA. Optimisation of deep-fat frying of plantain chips (Ipekere) using response surface methodology. Journal of Food Processing and Technology. 2016;**7**:1-6

[12] Ajiboye AO, Shodehinde SA. Diet supplemented with boiled unripe plantain (*Musa paradisiaca*) exhibited antidiabetic potentials in streptozotocin-induced Wistar rat. Journal of Food Biochemistry. 2022;**46**:1-9

[13] AOAC. Official Method of Analysis. Washington DC: Association of the Analytical Chemist; 2005

[14] Idowu AO. Nutritional, sensory and storage properties of snack produced from maize (*Zea mays* Linn) and African

yam bean seed (*Sphenostylis stenocarpa* Hochst Ex A. Rich) flour blends. [PhD dissertation]. Nigeria: University of Ibadan; 2014

[15] AACC, American Association of Cereal Chemists. Approved Methods of the AACC, Methods. 10th ed. St. Paul, MN, USA: Cereals & Grains Association; 2005

[16] Amoros W, Salas E, Hualla V, Burgos G, De Boeck B, Eyzaguirre R, et al. Heritability and genetic gains for iron and zinc concentration in diploid potato. Crop Science. 2020;**60**:1884-1896

[17] SAS. Statistical Analysis Software (SAS), Qualification Tools User's Guide SAS 9.2. Cary, NC, USA: SAS Institute Inc.; 2008

[18] Fadimu JF, Sanni LO, Adebowale AA, Kareem S, Sobukola PO, Kajihausa O, et al. Effect of drying methods on the chemical composition, color, functional and pasting properties of plantain (*Musa parasidiaca*) flour. Croatian Journal of Food Technology, Biotechnology and Nutrition. 2018;**13**(1–2):38-43

[19] Mba OG, Dumont MJ, Ngadi M. Influence of palm oil, canola oil and blends on characteristics of fried plantain crisps. British Food Journal. 2015;**117**(6):1793-1807

[20] Okole PA, Isirima CB, Ogunu-Ebiye UG, Chijioke-Eke JN. Proximate and mineral composition of plantain (*Musa paradisiaca*) flour obtained through different processing methods. Nigerian Agricultural Journal. 2022;**53**(1):137-142

[21] Okareh OT, Adeolu AT, Adepoju OT. Proximate and mineral composition of plantain (*Musa paradisiaca*) wastes flour: A potential nutrients sources in the formulation of animal feeds. African Journal of Food Science and Technology. 2015;**6**(2):53-57

[22] Omotosho OE, Garuba R, Ayoade F, Adebayo AH, Adedipe OE, Chinedu SO. Effect of deep fat drying using canola oil, soya oil and vegetable oil on the proximate, vitamins and mineral contents of unripe plantain (*Musa paradisiacal*). Journal of Applied Science. 2016;**16**(3):103-107

[23] Alamu EO, Ntawuruhunga P, Chileshe P, Olaniyan B, Mukuka I, Maziya-Dixon B. Nutritional quality of fritters produced from fresh cassava roots high-quality cassava and soy flour blends and consumer preferences. Cogent Food & Agriculture. 2019;**5**: 1677129

[24] Yarkwan B, Uvir RH. Effects of drying methods on the nutritional composition of unripe plantain flour. Food Science and Quality Management. 2015;**41**:5-10

[25] Ekesa B, Nabuuma D, Blomme G, Bergh I. Provitamin A carotenoid content of unripe and ripe banana cultivars for potential adoption in eastern Africa. Journal of Food Composition and Analysis. 2015;**43**:1-6

[26] Sojinu OS, Biliaminu NT, Mosaku AM, Makinde KO, Adeniji TH, Adeboye BM. The implication of ripening agents on chemical composition of plantain (*Musa paradisiaca*). Heliyon. 2021;**7**:1-5

[27] Inyang UE, Nkop SE, Umoh EB. Effect of the stage of ripening on the nutrients, anti-nutrients and functional properties of flours made from whole plantain fruit. Current Journal of Applied Science and Technology. 2017; **24**(1):1-9

Chapter 4

Environmentally Friendly Plant Terpenoids and Their Biological Activity

Salakhutdin Zakirov Khashimovich and Zulfiya Mukhidova Shabzalovna

Abstract

This chapter is devoted to the phytochemical study of sesquiterpene lactones of some plants of the Asteraceae family and their biological activity. At the same time, the effective growth activity of sesquiterpene lactones isolated by us from plants of the genus Artemisia was established, which improves the quality and increases the yield of rice, cotton and the productivity of mulberry cocoons. Also, a clear mutagenic, anti-nosema, antiviral and insecticidal activity of the amount of lactones isolated from plants of the genus Centaurea, Acroptilon and Handelia was revealed. A technology has been developed for obtaining biologically active compounds from plant materials and methods for the quantitative and qualitative determination of sesquiterpene lactones in plant materials and extracts.

Keywords: terpenoid, sesquiterpene lactone, Asteraceae, mutagen, sericulture, grena, cotton, rice, termite, extract

1. Introduction

The plant flora of Uzbekistan is very diverse and rich. From 8000 thousand species growing in Central Asia, it is represented by more than 4500 plant species. The basis of the flora of Uzbekistan is 10 large families: Asteraceae, Fabaceae, Poaceae, Brassicaceae, Rosaceae, Lamiaceae, Chenopodiaceae, Carophyllaceae, Liliaceae, Boraginaceae. At present, scientists of Uzbekistan have established groups of wild plants of natural flora for their use in medicine (600 species), food industry (400 species), essential oil (600 species), alkaloid-containing, glycoside-containing, coumarin-containing, flavonoid, terpenoid (3500 species), saponin-bearing 100 species, tanning 400 species, vitamin-bearing 600 species, fodder 1700 species.

Plants of family Asteraceae for wide application are easily available and widely distributed in the territory of our republic. Currently, medicine and agriculture mainly use synthetic preparations that are toxic and pollute the environment, while preparations based on plant materials do not have toxicity, but have a wide range of biological effects, and are environmentally friendly to humans, animals and the environment. However, in recent years, these plants have been severely exterminated and innovative research in this direction is not carried out to the full extent. In this regard, consistent reforms are being implemented in the republic for the protection of medicinal plants and their processing, conditions are being created to increase the export potential of the industry, as well as the integration of education, science and production processes.

In order to find and introduce new highly effective preparations and pesticides, it is necessary to expand the development of methods for obtaining biologically active compounds from wild and medicinal plants.

Sesquiterpene lactones, being a large group of secondary metabolites, are widely distributed in plants, and more than 5000 of their representatives with mono-, di- and tricyclic carbon skeletons have been isolated and established [1, 2].

2. Materials and methods of the research

The object of the research are widespread plants of the genus Artemisia, Centaurea and Handelia, family Asteraceae.

2.1 Research methods

Phytochemical, technological, physicochemical, biological.

Phytochemical and technological methods are used to develop the optimal technology for obtaining individual sesquiterpene lactones and the total lactones from plant materials (extraction of organic solvents, purification from ballast substances, separation by polarity and chromatographic separation, technological scheme for the isolation of biologically active compounds).

Physical and chemical methods - Infrared spectroscopy (IR), Ultraviolet spectroscopy (UFS), High performance liquid chromatography (HPLC) analyzes the completeness of the extraction of biologically active compounds (BAS) from plant materials, step-by-step control of technological processes of obtaining, determination of the qualitative and quantitative composition of biologically active substances (BAS) and identification of isolated compounds.

Biological methods determine the biostimulating, mutagenic, antiparasitic and insecticidal activity of the developed preparations.

2.2 Growth activity of rice growing and cotton growing

The objects of the study were widely distributed in Uzbekistan plants belonging to the Asteraceae family: Artemisia tenuisecta Nevski., Artemisia sogdiana Bunge., Artemisia leucodes Schrenk. and *Artemisia absinthium* L., producing active terpenoids, in particular sesquiterpene lactones α- and β-santonins, leukomizin, austricin, which have growth-stimulating activity.

Rice is the most common cereal crop in the world. It is a dietary product, has high nutritional properties, contains 70–80% starch and 7–8% protein, the latter being characterized by exceptionally high digestibility and a favorable balanced content of essential amino acids.

In terms of yield, rice ranks first among grain crops, and in terms of sown area and gross grain harvest, it ranks second.

To obtain high and stable yields of agricultural crops, it is necessary to introduce highly effective plant growth regulators that meet modern technology and environmental requirements. The need to use growth regulators in rice sowing is associated with low field germination of seeds, high empty grain of the panicle and lodging of the rice crop. There are various ways to regulate the growth and development of agricultural crops using natural and synthetic biologically active substances, such as auxins, gibberellins, cytokinins, etc. [3]. However, some of these substances are expensive, while others have not been widely used due to insufficient stability or relatively high toxicity (for example, auxins). Therefore, the task of creating and using cheap, non-toxic natural preparations that act, like phytohormones, in ultra-low concentrations, is relevant today. This task is also relevant for the cultivation of rice, since rice is one of the main food products of the inhabitants of the countries of Southeast Asia, including Uzbekistan, since the use of plant growth biostimulants during presowing seed treatment and spraying of rice crops makes it possible to increase the energy of seed germination, to obtain fast and friendly shoots, increase the development of the root system and plant biomass by 10%, leaf surface area and chlorophyll content, increase productivity.

The use of plant growth regulators with a versatile spectrum of action contributes to a significant reduction in the use of plant protection products against diseases and pests. Therefore, an integrated approach to the use of plant growth bioregulators, which have both growth-regulating and immunostimulating effects in the system of other technology elements, is still relevant at the present time.

To date, as a result of the studies, one of the effective biostimulants for increasing the yield of rice are natural sesquiterpenoids α-santonin, zerumbon and C_{16}-guayanolide, isolated from various plants, increase the yield of rice after one-day soaking of seeds in their solutions at a dilution of 1:10000 compared with control at 14.17% (santonin, zerumbon) and 7.5% (C_{16}-guayanolide) [4, 5].

α-santonin   zerumbon C_{16}-guayanolide

Our research in recent years found that the plants of the family Asteraceae of the flora of Uzbekistan are rich sources of biologically active sesquiterpenoids. For example, the aforementioned α-santonin, which is used in India as a biostimulant to increase the yield of rice, is produced in major quantities by plants of the genus Artemisia which are grown in Uzbekistan. These types of wormwood are the main edificators of plant communities in the arid and semi-arid zones of Uzbekistan and form wormwood pastures over a vast territory, and which can be used as a raw material for the production of α-santonin in the required quantities for its use as a growth stimulant.

We also considered the growth-regulating activity of a number of sesquiterpenoids liganolide, repin, granilin and artabine, including α-santonin, which we isolated from plants of the flora of Uzbekistan (**Figure 1**). Some of the results of these studies were published by the authors in the open press [6, 7].

An effective method of using plant growth and development regulators is the pre-sowing seed lock. The following methodology was used to consider the growth of stimulating activity. They took an exact weight of the terpenoid, dissolved it in a small amount of alcohol, and diluted it with warm water to a volume in a ratio of 1:10000.

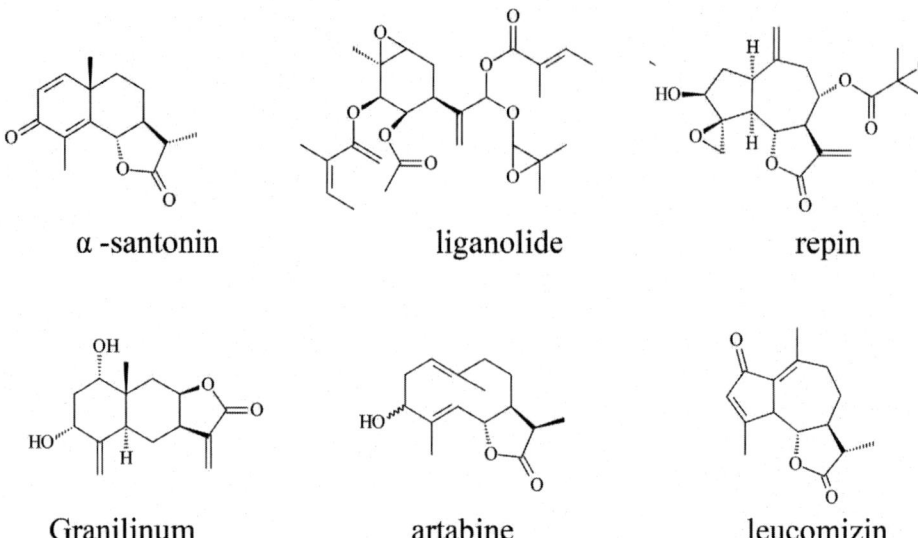

Figure 1.
Structural formulae of lactones with biostimulating activity.

Then, rice seeds of the Iskandar variety were immersed in the resulting solution of each terpenoid for 24 hours separately. The treated seeds were planted on experimental small plots row by row at a distance of 15 cm.

The yield was determined by dry weight and the results were compared with the control. As the results showed, the most active growth regulators were α-santonin, liganolide, repin, leucomizin, which significantly increased the yield of rice by an average of 12.5%, and granilin and artabine - up to 10%.

According to the list of pesticides and agrochemicals permitted for use in agriculture of the Republic of Uzbekistan, published in the collection (Tashkent, 2013), rice growth regulator "Edagum SM" (LLC "Spetsesnaska M Service", imported from Russia) is noted in our Republic used [8].

The main disadvantage of the Edagum SM biostimulant is the laboriousness of its use, which consists in the repeated processing of rice during the growing season to increase yields, which is not economically viable.

In this regard, we have been conducting research for a number of years in order to find and create a cost-effective, effective stimulant that increases the yield of rice.

As a result of ongoing research, a natural sesquiterpenoid 2-keto-8α-hydroxy-5α, 6β, 7α, 11β (H)-guai-1(10), 3-diene-6,10-olide monohydrate (1). The source of this sesquiterpenoid is wormwood from the subgenus Seriphidium, widely distributed in Uzbekistan.

The growth-regulating properties of this sesquiterpenoid, which we called the preparation "Risolid", can effectively increase the yield of rice with a single pre-sowing treatment of rice. Below are the results of biological studies of compound 1.

As can be seen from the data in **Table 1**, this biostimulant has a great influence on the germination of rice seeds.

Based on the analyzes, it can be seen that in the control variant, the growth of plants was 127 cm, the length of the panicle was 25.3 cm, the weight of the main panicle was 1.8 g, the weight of the lateral panicle was 0.8 g, and the yield was 55.9 centners per hectare. In the 2nd variant of the experiment (Edagum SM preparation), the plant growth was 132 cm, the length of the panicle was 27.8 cm, the weight of the main panicle was 1.8 cm, the weight of the lateral panicle was 0.9 g, the yield was

№	Experiment options	Number of germinated plants per 1 m² of area, pieces	Seed germination, %	Plant growth, cm
1	Control	256	51.2	10.5
2	Risolid 50 g/t seeds	281	56.2	11.5
3	Risolid 75 g/t seeds	279	55.9	11.2
4	Risolid 100 g/t seeds	280	55.7	10.8
5	125 g/t seeds	261	53.0	10.7
	Coef. Student R		2.7	0.05

Table 1.
Determination of seed germination in the field.

65.4 centners, which was 9.6 centners higher compared to control. In 3–4 variants of field experiments (Risolid preparation 50 and 75 g/t of seeds), plant growth was 135–136 cm, panicle length – 27.8–27.9 cm, main panicle weight 2.2 cm, lateral panicle weight 1 0–1.1 g, the yield was 70.3–70.4 centners, which was 14.4–14.5 centners higher compared to the control (see **Table 2**). Table 2 shows the rice yield indicators. Tashkent region, Srednechirchik district NIIR 2022.

The results of the tests indicate that the proposed preparation Risolid has a high growth-regulating activity at low concentrations, which leads to an increase in the yield of rice, and, thus, this agent can be used in agriculture to increase the yield of rice.

By studying the IR and HPLC of the isolated sesquiterpene lactones, their identification was carried out, the purity of the samples was established, and the quantitative and qualitative content of α-santonin, β-santonin, leukomisin and austricin in plant raw materials and extracts was determined using the HPLC method.

Ionizing radiation and chemical mutagens are mainly used in the selection of initial material in cotton growing. The use of mutagens of plant origin makes it possible to obtain valuable source material for breeding purposes in a short time.

Earlier, when studying the mutant effect of extractive substances *Artemisia absinthium* (preparation "PRP") on cotton, we obtained mutant lines. It was found that at various concentrations PRP has a pronounced biostimulating and mutagenic activity. Cotton seeds of varieties C-6524, C-6532, C-9070, etc. were treated with a preparation of various concentrations (2.0; 1.5; 1.0; 0.5%) and sown in the field. The following indicators were analyzed: seed germination, plant survival and the degree of development of traits characterizing productivity. In experimental variants at concentrations of 1.0 and 0.5%, there is a pronounced stimulation, in particular, the accumulation of fruit organs on a bush, an increase in the weight of raw cotton per plant, due to an increase in the number and weight of bolls [9–11].

This preparation at concentrations of 2 and 1% was used as a modifier to relieve depression during seed irradiation. For the study, the C-6524 variety was taken, as mutagenic factors—Co^{60} gamma rays and the PRP preparation at a concentration of 1 and 0.5%. Non-irradiated seeds, simply soaked in water, served as the control, and the second control was simply irradiated seeds.

As a result of the research, promising breeding material was obtained in the form of mutant lines and varieties. A number of promising mutant lines are being studied in the laboratory and nurseries ermacron mutagenesis. Two mutant lines are tested in competitive and station variety trials. Since the polar sum of wormwood showed biostimulating activity, in order to isolate the active principle, we studied the

№	Experiment options	Number of plants in 1 m², pieces	Bushing degree	Plant growth, cm	Panicle length, cm	Grain weight of one panicle, (average), g		Weight 1000 pcs. Grain, g	Biological yield, g	Real yield, c/ha
						Main panicle	Lateral panicle			
1	Control	240	1.12	127	25.3	1.8	0.8	28.9	698	55.9
2	Edagum SM	255	1.13	132	27.8	1.8	0.9	32.1	864	65.4
3	Risolid 50 g/t seeds	262	1.12	135	27.9	2.2	1.0	33.3	880	70.4
4	Risolid 75 g/t seeds	262	1.12	136	28.0	2.2	1.1	33.2	879	70.3
5	Risolid 100 g/t seeds	258	1.10	137	26.7	2.2	0.8	32.9	851	69.1
6	Risolid 125 g/t seeds	254	1.13	130	27.5	1.9	09	315	803	68.3

Note: Tashkent region, Srednechirchik district RRI 2022.

Table 2.
The results of field trials of the preparation Risolid in comparison with the prototype Edagum SM.

chemical composition of the aqueous extract of the aerial part of the raw material. A new water-soluble sesquiterpene lactone (I) of composition $C_{15}H_{22}O_5$ was isolated by chromatographic separation method, m.p.164-165°C. The IR spectrum acetyl derivative showed absorption bands at 3440 cm^{-1} characteristic of the tertiary hydroxyl group. The formation of this product confirms the presence of a secondary hydroxyl.

Oxidation of 1 with chromium (6) oxide gave keto derivative 3 with mp 183–184°C. The keto derivative 3 is identical in physicochemical constants and spectral data to the lactone artabsinolide B. Therefore, the new lactone 1 is an epimer of artabsinolide C [12]. Based on the above results, it can be concluded that the isolation and study of Artemisia absinthium terpenoids will lead to the creation of biostimulants for use in cotton growing.

2.3 Antinosema and antiviral activity in sericulture

On a global scale, the production of live cocoons is growing, along with this, due to dangerous infectious diseases, 15–20% of the cocoon crop is damaged. There is evidence that the sources of antibacterial, antiviral, antiprotozoal agents are medicinal plants containing various classes of natural compounds (alkaloids, quinones, polyphenols, saponins, terpenoids, sesquiterpene lactones) [13, 14].

The aim of our research was to test extracts (total sesquiterpene lactones) isolated from various plants against silkworm nosematosis.

After processing grena with agents representing the sum of sesquiterpene lactones *Artemisia tenuisecta* (ATEN), *Artemisia annua* (AANN), *Acroptilon repens* (AREP), *Centaurea squarrosa* (CESK) only in the variant 4, a decrease in the output of caterpillars was observed compared to the control by an average of 8.25%. In other variants, the revitalization of grena was higher than in the control and ranged from 81.5–89.0%. It is obvious that the revival of grena is to a certain extent dependent on the means used. The higher the concentration of the agent, the lower the percentage of caterpillar yield (**Table 3**).

Higher concentrations of solutions of the studied agents, as in medicine with pharmaceuticals, had a depressing effect on a living organism—silkworm grena. In cases using higher concentrations of extracts (2, 4), a slight increase in revitalization (ATEN 1.0–1.25%) or a serious decrease (AANN 1.0–8.25%) was recorded. In variants (1, 3, 5), where grena was treated in solutions of agents with a lower concentration, an increase in the yield of caterpillars (revitalization) by 4.50–8.75 absolute percent was observed. Under grena production conditions, this is a rather high percentage, indicating a significant increase in the quality of grena.

In the same experiment, another, much more important indicator of the quality of grena was taken into account—its infection with nosematosis, or rather, the decrease in infection under the influence of tested agents.

№ вар	Symbol of the agent and concentration,%	Grena revitalization after processing, %			In comparison with the control	
					Increase	Decrease
		$\bar{X} \pm S\bar{x}$	C_v	Pd	abs. %	abs. %
1	I (0.5%)	84.7 ± 0.20	0.4	0.999	4.50	—
2	I (1.0%)	81.5 ± 0.32	0.7	0.999	1.25	—
3	II (0.5%)	85.5 ± 0.45	0.4	0.999	5.25	—
4	II (1.0)	72.0 ± 0.32	0.8	0.999	—	8.25
5	IV (0.5%)	89.0 ± 0.76	1.5	0.999	8.75	—
6	III (0.5%)	81.5 ± 0.26	0.3	0.999	1.25	—
7	IV + III + I + II (0.5% each)	82.0 ± 0.08	0.1	0.999	1.75	—
8	Control (water)	80.25 ± 0.46	1.0			

Table 3.
Revitalization of grena after treatment with the amount of lactones of different concentrations.

№	Symbol of the agent and concentration,%	Infection of caterpillars-animators, %	Decrease in infection compared to control	
			abs. %	rel. %
1	I (0.5%)	4.5	3.6	44.4
2	I (1.0%)	1.8	6.3	77.8
3	II (0.5%)	3.2	4.9	60.5
4	II (1.0)	2.4	5.7	70.4
5	IV (0.5%)	0.9	7.2	88.9
6	III (0.5%)	2.6	5.5	67.9
7	IV + III + I + II (0.5% each)	2.4	5.7	70.4
8	Control (water)	8.1	—	—

Table 4.
Infection of caterpillars with nosematosis after treatment with the amount of sesquiterpene lactones of different concentrations.

Table 4 presents the results of infestation of caterpillars after grena treatment in solutions of sesquiterpene lactones.

Analysis of the obtained results presented in **Table 4** indicates that after the treatment of nosematosis-infected grena with solutions of the sum of sesquiterpene lactones, infection with nosematosis in hatched caterpillars decreased in all variants. Each caterpillar emerging from the treated grena was microscopically examined individually. Treatment with solutions of extracts of grena infected with nosematosis reduced the infection of emerging caterpillars by 3.6–7.2% absolute, or by 44.4–88.9 relative percent, compared with the control. The most effective, having a high anti-nositogenic effect, were agents with the code names CESK at 0.5% concentration, ATEN at 1.0% concentration and AANN at 1.0% concentration.

These funds are isolated from wormwood finely dissected, wormwood annual and cornflower splayed and have antihelminthic, antiprotozoal, antimalarial, as well as anti-inflammatory and immunostimulating properties. The causative agent of silkworm nosematosis, located in grena, belongs to the group of protozoal insect diseases. The results obtained by reducing the infection of caterpillars with nosematosis testify to the active effect of the amount of sesquiterpene lactones on the causative agent of the protozoan disease of the silkworm. An increase in the yield of caterpillars from grena treated with the above preparations indicates the immunostimulating effect of the tested agents.

The results obtained on the influence of the amount of sesquiterpene lactones on the infection of grena with nosematosis were processed by a mathematical method. Mathematical processing of the results of the experiment on the revitalization of grena and on the infection of revivalists caterpillars with nosematosis testifies to their reliability.

For clarity, we present the infection of revival caterpillars with silkworm nosematosis, hatched from grena, slightly infected with nosematosis and treated with herbal preparations (**Figure 2**).

As can be seen in **Figure 2**, the lowest infection with nosematosis is observed in revival caterpillars that emerged from a slightly infected grena treated with CESK (0.9%), ATEN 1.0 (1.8%), AANN 1.0 (2.4%). It should be noted that the use of plant biostimulants in high concentrations of ATEN 1.0 and AANN 1.0 caused a decrease in the revitalization of grena within the experiment, that is, it had a depressing effect on the development of the embryo (**Table 3**). The same terpenoids caused a decrease in the infestation of caterpillars-animators by 6.3 and 5.7 absolute or 77.8 and 70.4 relative percent, respectively (**Table 4**). Consequently, sesquiterpene lactones have a detrimental effect on silkworm nosematosis, while simultaneously suppressing biochemical processes in the body of the spore host.

Among the diseases of the silkworm, the most dangerous is also the viral disease nuclear polyhedrosis (jaundice). If this disease occurs on the rearing of the silkworm, then it is practically impossible to obtain any harvest of silk cocoons.

At present, there are no radical methods of combating this disease in Uzbekistan and neighboring countries, and sericulture suffers great material losses.

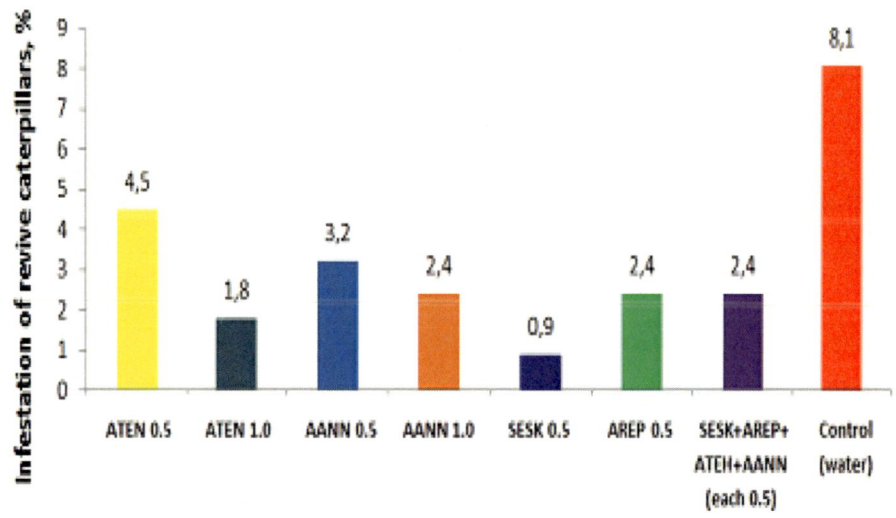

Figure 2.
Infection with nosematosis in revival caterpillars emerging from slightly infected and treated preparations of grena.

To solve this problem, in order to reduce the occurrence of spontaneous jaundice on silkworm rearings, terpenoids produced by the aerial part of Artemisia tenuisecta and A. leucodes are proposed, which have high antiviral activity and are able to inhibit the transition of the latent jaundice virus to the active state [15].

To be able to introduce funds based on sesquiterpenoids, the method of obtaining individual sesquiterpene lactones from plants of the flora of Uzbekistan, their subsequent alkaline hydrolysis is expensive, in this regard, the use of extractive amounts of sesquiterpene lactones converted into a water-soluble form by alkaline hydrolysis reduces the cost of the final product several times. In order to improve the quality of grena and the yield of mulberry cocoons, water-soluble extractive amounts of sesquiterpene lactones of three types of the aerial parts of plants of the flora of Uzbekistan were used: *Artemisia annua* L. (annual wormwood), Artemisia tenuisecta Nevski, Artemisia leucodes Schrenk (whitish wormwood).

The sum of sesquiterpenoids of the aerial part of Artemisia leucodes Schrenk contains: leukomizin (1), austricin (2), matrikarin (5), parishin B (6), parishin C (7), artelin (8), artelein (9), anhydroaustricin (10). It was previously established that the extractive sum of sesquiterpenoids, as well as individual lactones, exhibit antiprotozoal, choleretic and antiviral activities.

5: $R_1=R_3=H$, $R_2=OAc$

7: $R_1=R_2=H$, $R_3=OH$

8: $R_1=R_2=OH$, $R_3=H$

The aerial part of Artemisia tenuisecta Nevski contains mainly santonin (4), which is used as an anthelmintic.

4

The extractive sum of the aerial part of *Artemisia annua* L. contains sesquiterpene lactones arteannuin B (3), artemisinin (16), arteannuins H (17), K (18), L (19), M (20), 11,13-dihydroarteannuin B (21) and artemisinic acid (22) and exhibits bactericidal activity, inhibits the growth of the anthrax pathogen, the decoction is used as an anthelmintic, for respiratory infections, jaundice, skin diseases, malaria, tumors.

3

16   17   18   19   20

21   22

The above plants were collected during the period of budding and the beginning of flowering with the maximum content of sesquiterpene lactones in the Fergana and Tashkent regions.

As a result of the research, it was found that the created agent "ALEU" has biological activity that can inhibit the transition of the latent jaundice virus to the active state, that is, slow down the development of the disease. As a result of the treatment of grena with this preparation, there is an increase in revitalization by 3.4%, and viability increases by 5.8–8.0%. Accordingly, such caterpillars curl larger cocoons, with a mass of 0.11–0.15 g higher than that of caterpillars that emerged from untreated grain. The yield of cocoons from one box of grena increases by 3.6–4.1 kg.

Means "ALEU" is the amount of lactones isolated from plants of the domestic flora and is environmentally friendly to humans, animals and the environment.

The products of alkaline hydrolysis of sesquiterpene lactones leukomizin (1), austricin (2), arteannuin B (3), α-santonin (4) and their antiviral activity against the latent silkworm nuclear polyhedrosis virus were also studied.

Aqueous solutions of hydrolysis products of leucomizin, austricin, arteannuin B, and α-santonin were tested on silkworm grena, susceptible to latent nuclear polyhedrosis virus. The results of the experiment are presented in **Table 5**.

The data in the table show an increase in the revival of grena in most of the experimental options by 0.3–2.1% compared to the control option. In addition, in all experimental variants, the death of caterpillars from the induction of nuclear polyhedrosis, relative to the control, decreases and the viability of the silkworm increases after the induction of the nuclear polyhedrosis virus from 2.5 to 19.6%.

Thus, the tests carried out indicate a positive effect of the products of alkaline hydrolysis of sesquiterpene lactones leucomizin (1), austricin (2), arteannuin B (3) and α-santonin (4) on the revitalization of grena and the viability of the silkworm to the disease of nuclear polyhedrosis, which depends on the concentration of the solution and the structure of the studied lactones.

We have studied the effect of the extractive sum of sesquiterpene lactones (SPG) from the herb wormwood (*Artemisia absinthium*) on the productive and reproductive properties of the silkworm, since we have previously isolated biologically active sesquiterpene lactones from this plant. Testing aqueous solutions of the amount

Products of hydrolysis of lactones	Leucomizin (1)		Austricine (2)		Arteannuin B (3)		Santonin (4)		Water Control
Solution concentration (%)	0.5	0.1	0.5	0.1	0.5	0.1	0.5	0.1	—
The exit of the caterpillars from the processed grena (revitalization of the grena)	94.8	96.2	96.6	94.8	95.0	96.1	97.6	95.8	95.5
Total % death due to induction of the YP virus	27.1	30.7	32.8	23.8	28.9	35.5	30.2	32.0	43.4
Increase in viability compared to control (%)	16.3	12.7	10.6	19.6	14.5	7.9	13.2	11.3	—

Table 5.
Influence of alkaline hydrolysis products of sesquiterpene lactones on grena vitality and viability of silkworm.

of wormwood lactones (0.1–0.25%) leads to an increase in the productivity of the silkworm, which is expressed in an increase in silkiness by 3.1% and yield by 20%.

Thus, it becomes clear that when using biostimulants, one should be careful when choosing preparation concentrations in order to maintain a balance between the stimulating and inhibitory effects of sesquiterpene lactones.

2.4 Insecticidal activity

Currently, two types of termites Anacanthotermes turkestanicus Jacobs and Anacanthotermes ahngerianus Jacobs are widespread in Uzbekistan, which cause enormous damage to historical cultural monuments, buildings, structures and strategically important objects, present a special danger and a serious problem in the social and economic life of society. Although, large-scale work is underway to reduce the termite population, their distribution and the damage they cause is growing every year. Termites are social insects that can multiply quickly, endure the influence of extreme environmental factors and migrate from an unfavorable habitat to places more suitable for their habitat. Termites are able to preserve vitality and reproduce, even when part of the colony is torn off from primary reproductive individuals. Various chemicals used against termites have a short-term effect (3–5 days), and they also create problems associated with ecology and health, so the use of most of them is prohibited. In this regard, there is a need to develop new methods and means of controlling termites using poisonous food lures of intestinal prolonged action.

Recent studies have shown that cyclic sesquiterpenoids produced by plants of the Asteraceae and Apiaceae families are the most promising thermicidal preparations of intestinal prolonged action. For example, American researchers found that sesquiterpene lactone knicin with a germacrane type of skeleton isolated from *Centaurea maculosa* and sesquiterpene ketone vulgarone B from *Artemisia douglasiana* led to a high mortality rate among invasive termites. It was established that vulgarone B was lethal

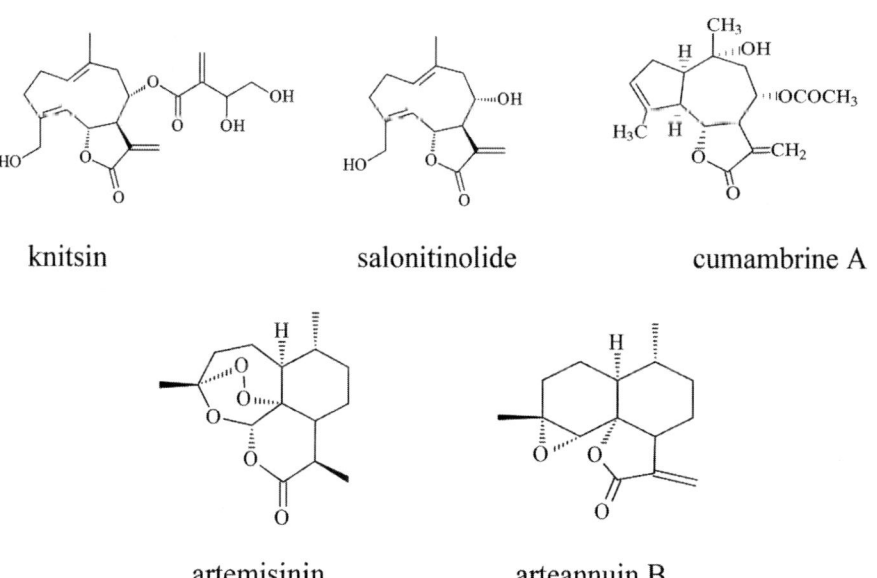

Figure 3.
Structural formulae of isolated Antitermitic sesquiterpene lactones.

Table 6.
Trial results of antitermite activity of sesquiterpene lactones artemisinin, knicin and kumambrin A.

Concentration	Number of termits	Observation day																			
		1		2		3		4		5		6		7		8		9		10	
		Π	Ж	Π	Ж	Π	Ж	Π	Ж	Π	Ж	Π	Ж	Π	Ж	Π	Ж	Π	Ж	Π	Ж
1. 0.001	20	—	20	—	20	1	19	—	19	2	17	—	17	2	15	5	10	7	3	3	0
2. 0.003	20	4	16	4	16	4	12	4	8	1	7	7	0								
3. 0.001	20	—	20	—	20	—	20	—	20	5	15	—	15	7	8	7	1	1	0		
4. 0.003	20	4	16	4	16	6	10	—	10	7	3	2	1	1	0						
5. 0.001	20	12	8	—	8	—	8	3	5	—	5	3	2	—	2	2	0				
6. 0.003	20	3	17	1	16	2	14	5	9	1	8	—	8	3	5	5	0				
Control	20	—	20	—	20	—	20	—	20	1	19	—	19	—	19	3	16	1	15	0	15

Note: Π—dead termites; Ж—living termites; 1, 2—artemisinin; 3, 4—knicin; 5, 6—kumambrin A.

for termites with a quick effect. On the fourth day after administration, vulgarone B showed a 97% mortality rate. Knicin acted more slowly, and on day 15 after completion he led to 81% termite mortality. Other authors found that the simultaneous use of several terpenoids with other additives leads to increased insecticidal activity. For example, a mixture of vulgarone B and knicine in low concentrations resulted in 96–100% termite mortality on day 15 after application [16, 17].

Our phytochemical studies established that the plants of the flora of Uzbekistan are rich sources of terpenoids. To date, as a result of joint research in Tashkent State agrarian University, Institute of the Chemistry of Plant Substances AS. Ruz, Institute of Zoology AS. Ruz, a number of effective antitermitic sesquiterpenoids of intestinal prolonged action from the domestic flora, such as knicin, kumambrin A, artemisinin, arteannuin B (**Figure 3**), which, when used individually, have been identified 96–100% termite death for 6–10 days of use (**Table 6**) [18, 19].

Using the HPLC method, a stage-by-stage control of the production of the technological process for obtaining thermicidally active sesquiterpene lactones salonitenolid and knicin from Centaurea squarroza and Jurinea maxima, kumambrin A from Handelia trichophylla, artemisinin and arteannuin B from *Artemisia annua* was developed.

3. Conclusion

A phytochemical study of sesquiterpene lactones of some plants of the Asteraceae family, followed by a study of their biological activity, has established a pronounced growth activity of sesquiterpene lactones isolated by us from plants of the genus Artemisia, which improves the quality and increases the yield of rice by 12–14 c/ha and increases the productivity of mulberry cocoons. On the basis of the total lactones of *Artemisia absinthium*, which has mutagenic activity, two new varieties of cotton with economically valuable traits have been created. Created anti-viral, anti-nosema and anti-termite agents based on the amount of lactones isolated from plants *Artemisia leucodes*, *A. annua*, *A. tenuisecta*, *Centaurea squarroza*, *Acroptilon repens* and *Handelia trichophylla* showed high antiviral activity to combat nuclear polyhedrosis and nosematosis of the silkworm, as well as brightly pronounced insecticidal activity, which destroys termites by 96–100%.

Using the HPLC method, a stage-by-stage control of the technological process for obtaining and determining the quantitative and qualitative composition of BAS in plant raw materials and extracts has been developed. Patents for the invention have been obtained for the growth and insecticidal activity of individual lactones.

Thus, the study of BAS flora of Uzbekistan is very relevant and promising, which will lead to the creation of highly effective environmentally friendly products for medicine and agriculture, the conservation of biodiversity and the rational use of local plant materials.

Author details

Salakhutdin Zakirov Khashimovich[1] and Zulfiya Mukhidova Shabzalovna[2*]

1 Department of Physics and Chemistry, Tashkent State Agrarian University, Tashkent, Uzbekistan

2 Department of General Ecology and Economy, Tashkent Branch of Astrakhan State Technical University, Tashkent, Uzbekistan

*Address all correspondence to: muxidova.zulfiya@mail.ru

IntechOpen

© 2023 The Author(s). Licensee IntechOpen. This chapter is distributed under the terms of the Creative Commons Attribution License (http://creativecommons.org/licenses/by/3.0), which permits unrestricted use, distribution, and reproduction in any medium, provided the original work is properly cited.

References

[1] Seaman FC. Sesquiterpene lactones as taxonomic characters in Asteraceae. The Botanical Review. 1982;**48**:121-594

[2] Merhatuly N. Chemistry of Mono- and Bicyclic Sesquiterpene γ-Lactones. Karaganda; 2015. p. 165

[3] Shevelukha RS. Plant Growth and its Regulation in Ontogenesis. M.: Kolos; 1992. p. 210

[4] Talwar KK, Kumar I, Kalsi PS. A dramatic role of terpenoids in increasing rice production. Experientia. 1983;**39**(1):117-119

[5] Talwar KK, Singh IP, Kalsi PS. A sesquiterpenoid with plant growth regulatory activity from *Saussurea lappa*. Phytochemistry. 1992;**31**(1):336-338

[6] Kh ZS, Sh MZ, DJ KK. Growth activity of terpenoids and their application in agriculture. "Vestnik" of the South-Kazakhstan State Pharmaceutical academy Republican Scientific Journal. 2014;**3**(68):78-79

[7] Kh ZS, Sh MZ, et al. Means for Pre-Sowing Treatment of Rice Seeds Patent for Invention. No. IAP 07090. Date of registration 30.09.2022; 2022

[8] List of Pesticides and Agrochemicals Permitted for Use in Agriculture of the Republic of Uzbekistan. Tashkent; 2017. 216 p

[9] Abdurakhmanov AA, Biyashev GZ, et al. Wormwood is a source of natural mutagens. Izv. AN. Kaz. SSR, biol. 1981;**6**:4-8

[10] Kovalchuk RI, Zakirov SK, et al. On the stimulating and mutagenic effect of extractive substances of wormwood on cotton. Uzbekistan Biological Journal. 1993;**6**:59-62

[11] Zakirov SK, Mukhidova ZS, Ibragimov PS. On the biological activity of plant terpenoids. In: IX International Scientific and Practical Conference "Agrarian Science for Agriculture". Barnaul; 2014. pp. 97-98

[12] Zakirov SK, Sham'yanov ID, Mukhidova ZS, Abdullaev ND. New watersoluble sesquiterpene lactone from *Artemisia absinthium* L. In: XIII International Symposium on the Chemistry of Natural Compounds. Shanghai, 16-19 October, 2019. 2019. p. 251

[13] Adekenov SM, Kulyyasov AT, Berdin AG, et al. Antiviral activity of sesquiterpene lactones. In: Tez. International Conference "HIV, AIDS and Related Problems" St. Petersburg. Vol. 3. 1999. p. 54

[14] Nurmurodova NF, Ismatullaeva DA, Mukhidova ZS, Zakirov SK. Effectiveness of natural terpenoids in the prevention of nosematosis of here. Solid State Technology Blind Peer Review Referred Journal. 2020;**63**:276-282

[15] Kashkarova LF, Shamyanov ID, Ziyaeva Ya M, Mukhamatkhanova RF. On the effect of sesquiterpene lactones on increasing the resistance of the silkworm to spontaneous nuclear polyhedrosis. Uzbekistan Biological Journal. 2004;**6**:12-16

[16] Tellez M, Osbrink W, Kobaisy M. Natural products as pesticidal agents for control of Formosan termite. Sociobiology. 2002;**6**

[17] Guillet C, Harmentha J, T.G. Waddell идр. Synergetic insecticidal

mode of action between sesquiterpene lactones and phototoxin, α- tertienyl. Photochemistry and Photobiology. 2000;**71**(2):111-115

[18] Kh ZS, Muxidova ZS. Natural ecologically safe antithermite agents. Journal of Science and Innovative Development. Tashkent. 2019;**N1**:73-77

[19] Kh ZS et al. Anti-Termite Agent and Method of Obtaining It. Patent for an invention. No. IAP 05623 Registration date 2018; 2018

Chapter 5

Cell Wall Enzymatic Activity Control: A Reliable Technique in the Fruit Ripening Process

Jamal Ayour, Hasnaâ Harrak and Mohamed Benichou

Abstract

The softening and structural changes that occur during fruit ripening are characteristic of specific species and can be attributed primarily to cell wall composition and cell swelling. Cell wall modifications are thought to result in changes in stiffness and texture, but the nature and extent of changes that occur during maturation vary widely. While some cell wall changes associated with ripening, such as depolymerization of matrix glycans, appear to be universal, other changes are highly variable in degree or present in different fruit types. However, the common point in all species is the involvement of the activities of enzymes linked to maturation in all these modifications, in particular the pectinolytic enzymes, namely polygalacturonase (PG), β-galactosidase (β-Gal) and pectin methyl esterase (SME). For good management of these changes, which have considerable consequences on the quality of fruits and their fate in post-harvest, the control of the activities of pectinolytic enzymes seems essential, which is what we propose to study in this chapter.

Keywords: texture, pectinolytic enzymes, quality, maturation, softening

1. Introduction

Ripening of fruits and vegetables is one of the final development stages of product ontogeny and involves many genetic, biochemical, and physiological changes. These changes include pigment and sugar accumulation, aromatic compound production, and meat tenderization [1]. These changes evolved to make fruits more attractive and edible to seed-dispersing organisms. Ripening also improves the sensory properties of the fruit, making it suitable for human consumption. However, once an advanced stage of ripening is reached, fruit quality deteriorates, mainly due to excessive fruit softening, increased susceptibility to pathogens, development of undesirable taste and skin color, etc., resulting in significant fruit management challenges and economic loss. Not only post-harvest longevity but also other economically important aspects determine yield [2]. Therefore, the rate of softening depends on handling procedures, harvest frequency, and the distance the fruit can be transported. Indeed, slowing down fruit softening is one of the main objectives and challenges of most product selection programs. This delay requires effective control of the factors involved in fruit ripening, especially softening.

2. Texture, a key criterion of ripening

Hardness and juiciness are the most important structural elements of fleshy fruits [3]. Both properties are primarily determined by parenchymal cell properties (cell wall thickness, shape, size, strength, and cell turgor pressure) and the extent and strength of adhesion zones between adjacent cells. During ripening, the parenchymal cell wall undergoes significant changes, altering its mechanical properties and greatly reducing cell adhesion due to the dissolution of the intermediate lamellae. Changes in the cell wall and intermediate lamellae that lead to fruit softening are usually caused by the activity of cell wall-modifying enzymes (e.g., polygalacturonase, pectin methylesterase, pectate lyase, β-galactosidase, and cellulase) encoded by ripening-related genes [4, 5]. Other cell wall proteins without hydrolase activity, such as B. expansin, also play a softening role [4]. In general, the cell wall degradation processes responsible for softening include depolymerization of matrix glycans, solubilization and depolymerization of pectin, and loss of neutral sugars from pectin side chains [5, 6]. The extent of these changes varies considerably between species. Recently, it has been suggested that the structural integrity of the xyloglucan network maintained by xyloglucosyltransferase/endohydrolase (XTH) may be important during fruit softening. This activity is usually higher during fruit development and then decreases or remains constant during ripening [5]. Miedes and Lorences [7] suggested that the XTH gene may be involved in the maintenance of cell wall structure rather than its degradation, and therefore, decreased expression and activity of the XTH gene may contribute to cell wall softening. This hypothesis is supported by the fact that overexpression of the SlXTH1 gene in tomatoes reduces fruit softening [7]. On the other hand, although less studied than cell wall degradation, cell swelling also affects fruit tenderness. During fruit ripening, a decrease in turgor pressure is often observed as the accumulation of dissolved apoplast is regulated. Water loss by transpiration through the cuticle may also be relevant, especially in fruits with thick, well-developed cuticles such as tomato [8]. Cell turgor pressure can also be influenced by cell wall changes that occur during fruit softening, so active changes in turgor pressure can be combined with passive moisture. It is difficult to distinguish from effects due to loss or changes in cell wall mechanical properties of fruit.

3. Main factors of softening during ripening

One of the main factors that reduce the quality of fruits and vegetables and cause significant economic losses is excessive softening. Changes in texture and changes during fruit ripening are mainly due to the dissolution of the interlayer due to the action of enzymes that modify the cell wall, the reduction of cell–cell adhesion, and the weakening of the parenchymal cell wall. Pectin, the main component of fruit cell walls, undergoes significant changes during ripening. These changes include solubilization, depolymerization, and loss of neutral side chains. Our work on apricot fruit [9, 10] and and recent evidence on strawberries [11] and apples [12, 13] characterized by a soft or crispy texture at maturity suggest that pectin disassembly is a key factor in texture changes during ripening. This change is mainly due to a commonality of active biomolecules, namely pectinolytic or pectic enzymes.

4. Pectinolytic enzymes

Pectinolytic enzymes or pectinases are a heterogeneous group of enzymes that hydrolyze pectic substances and are widespread in higher plants and microorganisms. This family of enzymes is able to attack a variety of chemical bonds in pectins. The term "pectinolytic enzyme" relates only to enzymes that act on the galacturonic part of pectic substances (**Figure 1**), and the enzymes capable of degrading the side chains are not classified among the pectolytic enzymes.

The enzymatic activity of the cell walls is linked above all to the pectolytic enzymes, which take part in the structural evolution of the wall by ensuring rearrangements or the degradation of the parietal polysaccharides and modifying and/or hydrolyzing the main components of the matrix. Different pectolytic enzymes are involved in these modifications. Pectinases are involved in the modification of parietal polysaccharides. For the majority of fruits, most of the cell wall changes are explained by the increased enzymatic activities of polygalacturonase (PG), β-galactosidase (β-Gal), and pectin methyl esterase (PME) [10, 15, 16].

4.1 Polygalacturonase

Polygalacturonases (PG) are glycosidases that hydrolyze the α-(1 → 4) glycosidic bond between unesterified galacturonic acid residues and in homogalacturonic chains (**Figure 1**). Recent works [16, 17] have shown differences in PG activity in different

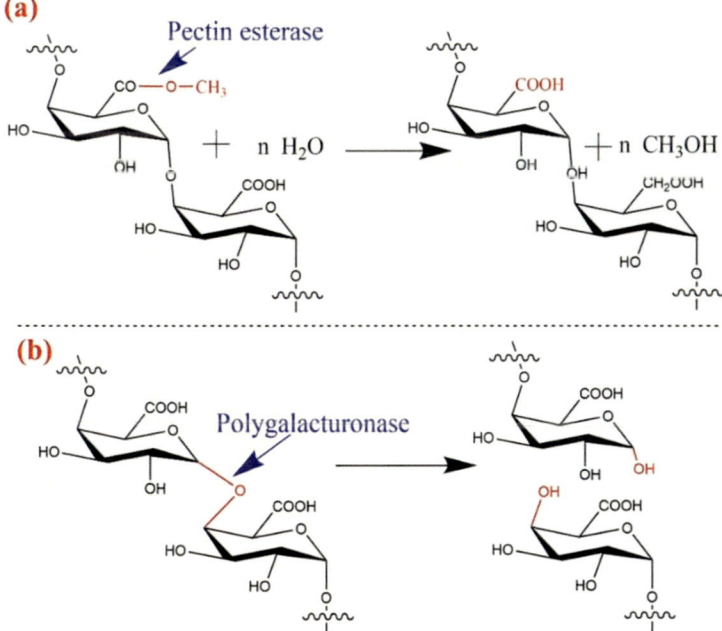

Figure 1.
Pectinase interaction model: (a) pectin esterase interaction model, (b) polygalacturonase [14].

varieties of apricot but with low levels compared to the activities found in other fruits and vegetables [18]. Cardarelli et al. [19] did not detect polygalacturonase activity in apricots. Endo-polygalacturonase (endoPG) activity has been associated with loss of firmness in several species, but this has never been confirmed in apricots. In tomatoes, the observation of a strong endo-PG activity in ripe fruits showed that PGs play an important role in the loss of firmness. For peaches, the activity of polygalacturonases in different cultivars is positively correlated with the loss of firmness and is only induced when the fruit ripens.

Two methods have been developed to determine the activity of PGs. This activity can be monitored by measuring the decrease in viscosity or the increase in the reducing power of the substrate (pectic acid or pectin). The comparison of viscosity and reducing power measurements during the depolymerization of pectins and pectic acids makes it possible to distinguish between "endo" and "exo" PG activities.

4.2 Pectin methylesterase

Pectin methylesterase (PME, EC 3.1.1.11) is an enzyme ubiquitous inside the plant kingdom. However, its function in plant increase and improvement continues to be unclear. Pectin methylesterase (PME) is involved in the loss of firmness by demethylating pectin (**Figure 1**), making it sensitive to the activity of PGs. Ünal and Şener [20] studied the biochemical properties of MSY in Alyanak apricots (an important variety in the Malatya region of Turkey). This PME has a high activity for a pH between 7.0 and 8.0 with a maximum activity at pH 7.5, knowing that the pH of an apricot fruit for example generally varies between 3.3 and 4.0. The enzyme is stable at temperatures between 30 and 40°C for 10 min and loses all its activity after 10 min at 80°C, which implies that PME can easily be inactivated by a pasteurization process during the processing of mumps apricots in syrup or jam. Other recent studies have reported PME activity in apricot fruit [10, 16, 17, 19, 21].

The activity of the SME is influenced by several factors, in particular, the stage of maturity of the fruits. PG and PME activities increase significantly ($p < 0.05$) with maturity (green, green mature, and ripe) [17]. In all the apricot cultivars studied, the PG activity increased from 713 to 14,286 nkat mg^{-1}. The highest PME activity was found at the ripe stage: green (45 nkat mg^{-1}) and ripe (128 nkat mg^{-1}) fruits. However, Ribas-Agusti et al. [16] reported that MSY activity tends to decrease with ripening for the studied apricot varieties.

The activity of PE can be monitored either by assaying the methanol released, or by determining the increase in the number of free carboxyls, or even by using a pH regulator. Indeed, the ionization of the carboxyl group produces a proton in the medium, which causes a variation in pH. PE is inhibited by the increase in the number of free carboxyls along the progressively demethylated polygalacturonic chains. This inhibition is due to the repulsion exerted by the negative charge of the ionized carboxyls. The presence of cations (Ca^{2+} and $Na+$) could counteract this inhibition. This inhibition of PEs would also be due to the side chains of neutral sugars in the pectin molecule [22].

4.3 β: D-galactosidase

The hydrolysis of cell wall pectins during maturation is also linked to the activity of glycosidases and especially β-D-Galactosidase which hydrolyzes galactosyl polymers. β-galactosidase leads to the loss of galactose units from the side chains [23]. Galactose was found as a product of this activity during tomato ripening. Kovács and

Szerdahelyi [24] reported that the stage of maturity greatly influences the activity of apricot galactosidases. Ribas-Agusti et al. [16] also reported that β-Gal activity tended to increase during maturation for all cultivars analyzed. Other work has shown that the solubilization of pectins by β-galactosidase accelerates the softening of fruits in general [4] and melon in particular.

5. Ripening management and enzymatic activity control

Recent studies [10, 12, 13, 16, 25] have shown that the degradation of polysaccharides in the cell wall results in A synergistic effect occurs between several enzymes that modify the cell wall, including PME, PG, and β-GAL. Consistent with our study and other reports [26, 27], apricot fruit wall enzyme activity increased during the fruit ripening process. Indeed, these enzymatic activities represent a major asset for managing the ripening of fruits and vegetables and determining the stages of development, the optimal stage of harvesting, and post-harvest management.

Fan et al. [25], recently, reported that the use of suppressed NFT storage can inhibit the enzymatic activities. Indeed, the PME activity of apricots increased rapidly from the beginning of storage at 5°C, but this increase in enzyme activity was effectively inhibited by storage at 0°C and NFT. Compared to storage at 0°C, NFT

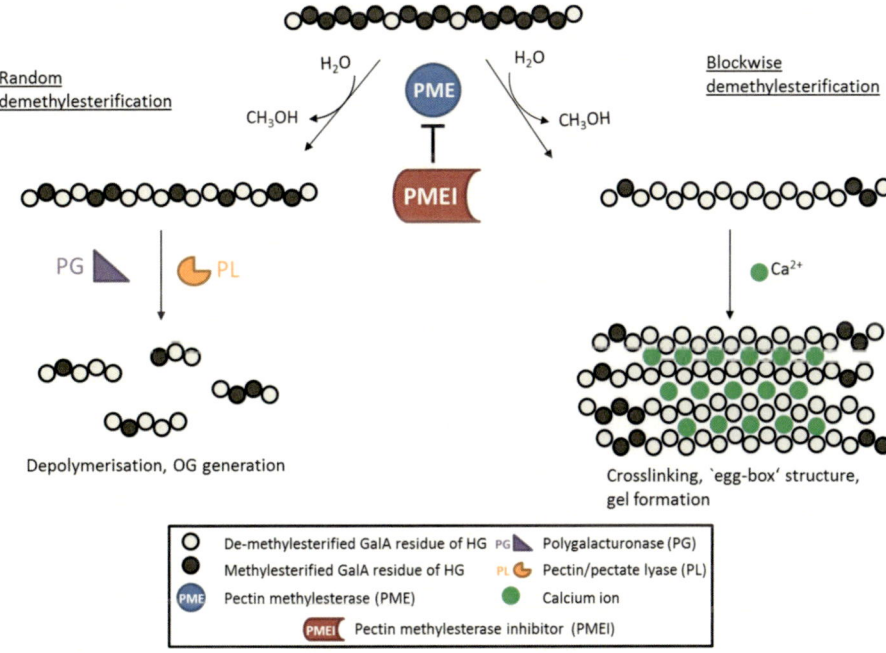

Figure 2.
Schematic diagram of HG demethylesterification and its effect on structure. HG is highly methylesterified when deposited on the cell wall. PME can demethylate HG in blocks, resulting in multiple contiguous GalA residues without methyl ester groups. Because these HG scaffolds are negatively charged, they can cross-link with cations such as calcium ions, resulting in the formation of so-called "egg-crate" structures that are responsible for gel formation. On the other hand, PME can demethylate individual HisGalA residues, causing random methyl esterification patterns. Low-level HG methyl ester is depolymerized by pectinolytic enzymes such as pectin/pectate lyase (PL) and polygalacturonase (PG), leading to the formation of oligogalacturonides (OG). In contrast, PME activity is inhibited by the protein inhibitor PMEI.

storage reduced the enzyme activity of apricots to a lower level. The PME activity of apricots stored in NFT was 88% of that of apricots stored at 0°C on day 60.

Moreover, the changes in apricot enzyme activities showed similar trends for all cell wall enzymes.

Genetically, in fruits, softening was reduced due to the antisense downregulation of polygalacturonase genes [28]. Indeed, changes in pectic polymer size, composition, and structure have traditionally been studied by conventional techniques; other studies focusing on changes at the nanostructural and genetic level have reported that gene regulation of enzymes is a solution for a better management of their activity during maturation and indeed consequences on the harvest [29].

Pectin methylesterase (PME) activity is controlled by a family of protein inhibitors called pectin methylesterase inhibitors (PMEIs) (**Figure 2**). Therefore, the interaction of PME and PMEI is considered not only as a determinant of cell adhesion, cell wall porosity and elasticity, but also as a source of release of signaling molecules during cell wall stress during fruit development stages. Wormit and Usadel [29], highlighted the importance of the PMEI gene family, its regulation and structure, its interaction with PMEI, and its function in response to stress during fruit and vegetable development and crop management.

6. Conclusion

Textural changes in fruits during ripening are a result of cell wall changes implied by enzymatic activity. These modifications have considerable consequences on the quality of the fruits and their fate after harvest; in fact, controlling the activities of pectinolytic enzymes is a key to managing the harvest and optimizing fruit quality.

Indeed, control tools have been developed and tested, controlling the activity of pectinolytic enzymes, specifically PME, PG, and β-GAL, during ripening is a preventive measure in the management of the fruit harvest. The choice of fruit storage conditions after harvest, such as NFT storage, is a solution that makes it possible to inhibit the activity of enzymes. Other studies are underway on the genetic regulation of pectinolytic enzymes to manage and modify their activity. This will make it possible to optimize the quality of the fruits, both for the consumer market and also for industrial processing.

Author details

Jamal Ayour[1*], Hasnaâ Harrak[2] and Mohamed Benichou[3]

1 Bioprocess Laboratory (LASIME), Higher School of Technology and National School of Applied Sciences, Ibn Zohr University, Agadir, Morocco

2 Laboratory of Agri-Food Technology and Quality, Regional Center for Agricultural Research in Marrakesh, National Institute for Agricultural Research (INRA), Marrakesh, Morocco

3 Food Sciences Laboratory, Faculty of Sciences Semlalia, Cadi Ayyad University, Marrakesh, Morocco

*Address all correspondence to: jamayour@gmail.com

IntechOpen

© 2023 The Author(s). Licensee IntechOpen. This chapter is distributed under the terms of the Creative Commons Attribution License (http://creativecommons.org/licenses/by/3.0), which permits unrestricted use, distribution, and reproduction in any medium, provided the original work is properly cited. [(cc) BY]

References

[1] Ayour J, Sagar M, Harrak H, Alahyane A, Benichou M. Evolution of some fruit quality criteria during ripening of twelve new Moroccan apricot clones (Prunus armeniaca L.). Scientia Horticulturae. 2017;**215**:72-79

[2] Bapat VA, Trivedi PK, Ghosh A, Sane VA, Ganapathi TR, Nath P. Ripening of fleshy fruit: Molecular insight and the role of ethylene. Biotechnology Advances. 2010;**28**:94-107

[3] Toivonen PMA, Brummell DA. Biochemical bases of appearance and texture changes in fresh-cut fruit and vegetables. Postharvest Biology and Technology. 2008;**48**(1):1-14

[4] Brummell DA, Harpster MH. Cell wall metabolism in fruit softening and quality and its manipulation in transgenic plants. Plant Molecular Biology. 2001a;**47**:311-340. DOI: 10.1023/A:1010656104304

[5] Goulao LF, Oliveira CM. Cell wall modifications during fruit ripening: When a fruit is not the fruit. Trends in Food Science and Technology. 2008;**19**:4-25

[6] Brummell DA. Cell wall disassembly in ripening fruit. Functional Plant Biology. 2006;**33**:103-119

[7] Miedes E, Lorences EP. Xyloglucan endotransglucosylase/hydrolases (XTHs) during tomato fruit growth and ripening. Journal of Plant Physiology. 2009;**166**(5):489-498. DOI: 10.1016/j.jplph.2008.07.003. Epub 2008 Sep 11

[8] Saladié M, Matas AJ, Isaacson T, Jenks MA, Goodwin SM, Niklas KJ, et al. A reevaluation of the key factors that influence tomato fruit softening and integrity. Plant Physiology. 2007;**144**(2):1012-1028

[9] Ayour J, Gouble B, Reling P, Ribas-Agusti A, Audergon JM, Maingonnat JF, et al. Impact of cooking on apricot texture as a function of cultivar and maturity. LWT. Food Science and Technology. 2016;**85**:385-389

[10] Ayour J, Le Bourvellec C, Gouble B, Audergon JM, Benichou M, Renard CM. Changes in cell wall neutral sugar composition related to pectinolytic enzyme activities and intra-flesh textural property during ripening of ten apricot clones. Food Chemistry. 2021;**339**:128096

[11] Kim D, Jeon SJ, Yanders S, Park S, Kim HS, Kim S. MYB3 plays an important role in lignin and anthocyanin biosynthesis under salt stress condition in arabidopsis. Plant Cell Reports. 2023;**41**:1549-1560. DOI: 10.1007/s00299-022-02878-7

[12] Huang Q, Lin B, Cao Y, Zhang Y, Song H, Huang C. CRISPR/Cas9-mediated mutagenesis of the susceptibility gene OsHPP04 in rice confers enhanced resistance to rice root-knot nematode. Frontiers in Plant Science. 2023;**14**:1134653

[13] Li Y, He H, Hou Y, Kelimu A, Wu F, Zhao Y, et al. Salicylic acid treatment delays apricot (*Prunus armeniaca* L.) fruit softening by inhibiting ethylene biosynthesis and cell wall degradation. Scientia Horticulturae. 2023;**300**:111061

[14] Polizeli Mde L, Jorge JA, Terenzi HF. Pectinase production by Neurospora crassa: Purification and biochemical characterization of extracellular polygalacturonase activity. Journal of General Microbiology. 1991;**137**(8):1815-1823. DOI: 10.1099/00221287-137-8-1815

[15] Kurz C, Carle R, Schieber A. Characterisation of cell wall

polysaccharide profiles of apricots (*Prunus armeniaca* L.), peaches (*Prunus persica* L.), and pumpkins (*Cucurbita* sp.) for the evaluation of fruit product authenticity. Food Chemistry. 2008;**106**:421-430

[16] Ribas-Agusti A, Gouble B, Bureau S, Maingonnat JF, Audergon JM, Renard CMGC. Towards the use of biochemical indicators in the raw fruit for improved texture of pasteurized apricots. Food and Bioprocess Technology. 2017;**10**:662-673

[17] Abaci ZT, Asma BM. Changes in some enzymatic parameters of six apricot cultivars during ripening. Anadolu Journal of Agricultural Sciences. 2014;**29**:174-178

[18] Nunes C, Saraiva JA, Coimbra MA. Effect of candying on cell wall polysaccharides of plums (*Prunus domestica* L.) and influence of cell wall enzymes. Food Chemistry. 2008;**111**:538-548

[19] Cardarelli M, Botondi R, Vizovitis K, Mencarelli F. Effects of exogenous propylene on softening, glycosidase, and pectinmethylesterase activity during postharvest ripening of apricots. Journal of Agriculture and Food Chemistry. 2002;**50**:1441-1446

[20] Ünal MÜ, Şener A. Extraction and characterization of pectin methylesterase from Alyanak apricot (*Prunus armeniaca* L). Journal of Food Science and Technology. 2015;**52**(2):1194-1199

[21] Botondi R, DeSantis D, Bellincontro A, Vizovitis K, Mencarelli F. Influence of ethylene inhibition by 1-methylcyclopropene on apricot quality, volatile production, and glycosidase activity of low- and high-aroma varieties of apricots. Journal of Agriculture and Food Chemistry. 2003;**51**:1189-1200

[22] Sakai T, Sakamoto T, Hallaert J, Vandamme EJ. Pectin, pectinase and protopectinase: Production, properties and applications. Advances in Applied Microbiology. 1993;**39**:213-279

[23] Barnavon L, Doco T, Terrier N, Ageorges A, Romieu C, Pellerin P. Analysis of cell wall neutral sugar composition, β-galactosidase activity and a related cDNA clone throughout the development of Vitis vinifera grape berries. Plant Physiology and Biochemistry. 2000;**38**:289-300

[24] Kovács E, Németh-Szerdahelyi E. β-Galactosidase activity and Cell Wall breakdown in apricots. Journal of Food Science. 2002;**67**:6

[25] Fan X, Jiang W, Gong H, Yang Y, Zhang A, Liu H, et al. Cell wall polysaccharides degradation and ultrastructure modification of apricot during storage at a near freezing temperature. Food Chemistry. 2019;**300**:125194

[26] Carvajal F, Palma F, Jamilena M, Garrido D. Cell wall metabolism and chilling injury during postharvest cold storage in zucchini fruit. Postharvest Biology and Technology. 2015;**108**:68-77

[27] Liu K, Liu J, Li H, Yuan C, Zhong J, Chen Y. Influence of postharvest citric acid and chitosan coating treatment on ripening attributes and expression of cell wall related genes in cherimoya (*Annona cherimola* mill.) fruit. Scientia Horticulturae. 2016;**198**:1-11

[28] Senechal F, Wattier C, Rusterucci C, Pelloux J. Homogalacturonan-modifying enzymes: Structure, expression, and

roles in plants. Journal of Experimental Botany. 2014;**65**:5125-5160

[29] Wormit A, Usadel B. The multifaceted role of pectin Methylesterase inhibitors (PMEIs). International Journal of Molecular Sciences. 2018;**19**(10):2878

Chapter 6

Artificial Ripening Technologies for Dates

Maged Mohammed, Nashi K. Alqahtani and Muhammad Munir

Abstract

Date palm fruits have essential importance due to their high economic value, nutritional benefits, and contribution to food security in arid and semi-arid regions. The unfavorable climatic conditions, drought or water scarcity, inconsistent pollination, genetic factors, and nutrient deficiencies cause date fruits to remain unripe for a long time. Artificial ripening is hastening fruit ripening using various techniques and chemicals. Artificial ripening techniques are employed to ripen date palm fruits to reduce their spoilage and waste, enhance their quality, and extend their shelf life. Therefore, artificial ripening has an economic benefit by supplying high-quality fruit, potentially increasing farmers' profits. However, using safe and approved techniques for artificial ripening is essential, as some processes can have negative health influences if misused. This chapter aims to discuss the concept of artificial ripening for date palm fruits and its benefits, explore various chemical and physical methods, analyze their effects on fruit quality, and examine the regulatory and safety considerations associated with artificial ripening. Additionally, the chapter examines the advantages and disadvantages of different ripening methods and their corresponding effects on the dates' nutritional value and sensory quality. The chapter highlights the need for sustainable and safe artificial ripening practices to meet consumer demand and ensure the high quality and availability of date palm fruits.

Keywords: date palm, fruit ripening, climate change, food safety, quality improvement, food sustainability, physicochemical characteristics

1. Introduction

The date palm (*Phoenix dactylifera* L.), a cross-pollinated fruit tree, is cultivated in arid and semi-arid regions for its sweet fruit and a staple food in dry regions. With over 120 million trees globally, it is primarily grown in the Arab world, with the majority in Egypt, Iraq, Saudi Arabia, Algeria, Morocco, Tunisia, and the UAE. Arab countries have 70% of the world's date palm trees, contributing 67% to global production. It is native to North Africa and the Middle East. It is one of the most valuable crops in these regions, and people use its fruits, leaves, and sap for medicine, food, and shelter [1]. It is cultivated on 1.31 million hectares and produces 9.82 million tons of fruit annually worldwide [2]. Its importance can be observed in several domains, including agriculture, nutrition, economic nutrition, traditional medicine, and culture. Date palm cultivation has significantly influenced the growth of oasis agriculture systems, providing shade and microclimates for crop growth. These trees also prevent soil erosion due to their extensive root systems. The date palm fruit is a sweet, fleshy berry rich in calories, sugars, fiber, vitamins A, B1, B2, and C, and minerals such as potassium, calcium, and iron. The fruit is also a good energy source and can be consumed fresh, dried, or processed into syrup, paste, and vinegar. Date palm sap, a sweet, milky liquid extracted from the tree trunk, makes date honey syrup, alcoholic beverages, and vinegar. The global date market, worth billions of dollars annually, includes fresh dates, processed products like date syrup, date paste, and date-based sweets, contributing to the economy [3, 4].

The cultivation of date palms is significant for several reasons, including its economic importance. The date palm industry is labor-intensive and helps both males and females by generating revenue and jobs. Due to increased employment prospects in rural areas, widespread migration to cities is lessened. Women have a significant role, especially during the palm propagation and post-harvest stages, including packaging and marketing [5]. Date production and trading help local economies and serve as a source of revenue for farmers and exporters. Over the past few decades, the Kingdom of Saudi Arabia's agricultural sector in general and the date palm sector, in particular, have experienced tremendous growth and support [6]. Many countries in the Middle East and North Africa, such as the Kingdom of Saudi Arabia, Iran, Iraq, Egypt, and Tunisia, rely substantially on the export of dates. In 2021, the Kingdom of Saudi Arabia was the leading global exporter of fresh or dried dates, with an export value of about 322.84 million USD followed by Israel (317.07 million USD), Iran (305.23 million USD), Tunisia (255.90 million USD), United Arab Emirates (209.43 million USD), Algeria (140.79 million USD), and United States of America (116.79 million USD) [7].

The cultural value of date palm cultivation contributes to its significance. For thousands of years, dates have been integral to the Middle East and North Africa's food and culture. They are frequently offered at special occasions and festivals and are mentioned in many religious writings. In many civilizations, date palms are regarded as symbols of hospitality and prosperity [5, 8]. Date fruits are believed to have laxative properties, aiding in constipation relief. Seeds are used in poultices for skin conditions like burns and wounds. The leaves and sap of the tree are used in traditional remedies for fever, diarrhea, and respiratory disorders. The cultivation of date palms is also important for the ecosystem. They are one of the valuable crops in areas with scarce water resources. Stabilizing sand dunes with their extensive roots can also help prevent soil erosion and desertification [9, 10]. In addition to their economic, cultural, and environmental significance, dates also have nutritional benefits. They are rich in carbohydrates, vitamins, fiber, and minerals. In traditional medicine, dates have been used for medicinal purposes [4, 11].

The unfavorable environmental conditions, climate change, pest and disease infestation, inconsistent pollination, genetic factors, nutrient deficiencies, and droughts affect the natural ripening process of date palm fruits. These unfavorable parameters can result in some fruits remaining unripe while others ripen or disrupt the natural ripening process. Therefore, date palm fruits do not naturally ripen simultaneously, leading to a prolonged ripening process and causing uneven ripening within a bunch [12, 13]. Artificial ripening techniques are employed to hasten fruit ripening using various techniques and chemicals. Artificial ripening techniques for date palm fruits offer benefits such as consistent supply, increased marketability, extended shelf life, control over the ripening process, reduction in post-harvest losses, access to distant markets, and environmental sustainability. This chapter aims to highlight the applications of artificial ripening in ensuring a consistent supply of ripe fruits, improving fruit quality and shelf life, and reducing post-harvest losses. Additionally, it discusses the potential impact of artificial ripening on the nutritional value of date palm fruits. It provides an overview of the global regulations and standards governing the artificial ripening process.

The rest of this chapter is structured as follows: First, we introduce the impact of environmental change on date ripening in Section 2; Section 3 describes artificial ripening and its benefits; Section 4 provides an overview of chemical methods of artificial ripening using ethylene (C_2H_4), calcium carbide (CaC_2), ethanol (C_2H_6O), and ethephon ($C_2H_6ClO_3P$); Section 5 provides benefits and application of physical methods of artificial ripening using hot air circulation, steam treatment, radiation exposure, and application of solar energy for artificial fruit ripening; Section 6 details effects of artificial ripening on date quality. Section 7 indicates the regulatory and safety considerations; Section 8 concludes the work.

2. Impact of environmental change on dates ripening

Environmental conditions such as temperature, humidity, rainfall, and other factors can significantly influence plant growth and development. Climate change is a significant concern for socioeconomic activities, including agriculture. Over the past 30 years, its effects on plant growth and development have become evident. Global evidence suggests climate change is occurring, with land surface temperatures increased by 0.6 ± 0.2°C over the twentieth century and predicted to rise by 1.4–5.8°C by 2100 [14, 15]. Climate change is altering global seasonal rainfall patterns, affecting crop responses due to rising temperatures, changing water resources, and rising carbon dioxide concentrations. It is being monitored in various physical and biological systems, with plant phenology being a crucial bio-indicator due to its ability to provide significant temporal and spatial information regarding ongoing changes [16]. This negatively impacts plant flowering and fruiting patterns, leading to potential shifts in climate suitability for date palm farming. The changing climate is expected to alter the suitability of certain regions for date palm farming, while others that are currently unsuitable may become suitable in the future [17].

Temperature is crucial in controlling plant reproductive phenology, particularly in tree species. Climate change has significantly impacted global weather patterns, causing increased temperatures and decreased rainfall. This impact is particularly severe in arid and semi-arid dry regions where date palm is commonly grown as a staple food. Date palm requires a long summer with high temperatures, a mild winter without frost, no rain during flowering and fruit set, low humidity, and plenty of sunshine [17, 18]. Temperature, humidity, and precipitation are some environmental factors affecting the ripening process of date palm fruit. Date palm fruit ripening can be affected by climate change directly or indirectly, depending on how these variables change. Climate change impacts date palm fruit ripening through temperature patterns. High temperatures during the growing season can accelerate ripening, causing premature fruit drops and reduced quality, while low temperatures can delay ripening and prolong fruit maturity. These temperature conditions are crucial for optimal growth and fruit development in date palms. Precipitation patterns significantly impact date palm fruit ripening. Despite their arid adaptation, date palms require water for growth and development [13, 19]. Climate change-related changes, like droughts or heavy rainfall, can negatively affect date palm fruit ripening. Drought conditions can cause water stress, affecting fruit size and quality. Excessive rainfall can damage fruits, promote fungal diseases, and reduce insect activity, affecting pollination and fruit set. Humidity levels significantly impact date palm fruit ripening, with high humidity increasing the risk of fungal diseases and pollination, while low humidity can lead to excessive fruit moisture loss, resulting in shriveled and poor-quality dates [20, 21]. Climate change can disrupt the natural ripening process of date palm fruits by altering day length (degree days) through changes in cloud cover or atmospheric conditions, as date palm is sensitive to these changes.

Studies suggest that the number of heating degree days impacts fruit softening. The finest date cultivars require 3300 heat units for full maturity; however, this varies by country, region, and cultivar [18]. In addition, climate change can indirectly affect date palm fruit ripening by altering pest and disease dynamics. Rising temperatures and altered precipitation patterns can create more favorable conditions for pests and diseases, leading to fruit damage and reduced fruit quality. Climate change has caused a delay in the *in situ* fruit ripening of date palms in Saudi Arabia over the past few years. The outer spikelets of fruits remained unripe, while the inner spikelets ripened. The irregular ripening is believed to be caused by a temperature and humidity imbalance (per. Comm.).

3. Artificial ripening and its benefits

Fruits supply essential nutrients that the body needs to maintain health. They are mostly consumed at mature or ripening stages. Because of their irregular *in-situ* ripening, climacteric fruits are harvested at a mature stage and artificially ripened afterward. Many date palm cultivars are not edible due to soluble tannins at the full coloration stage (Khalal/Bisr). Date growers face prolonged Rutab development, especially when ripening is uneven, exposing fruits to abscission. The beginning of the Rutab stage is defined as the initiation of ripening in date fruits, which proceed to the Tamr or full ripe stage. Date growers must wait for 50% of the bunch to ripen, and a large amount must be harvested quickly, necessitating the development of an artificial ripening method at the Khalal/Bisr stage. Artificial ripening is the process of hastening fruit ripening with various techniques and chemicals. It is commonly used in the agricultural industry to ensure a steady supply of ripe fruits year-round despite seasonal variations. It is a controlled ripening process to enhance consumer acceptance and sales by achieving desired characteristics [22, 23]. Artificial fruit ripening techniques such as using chemicals (calcium carbide, ethylene, ethanol, ethephon, acetylene gas, and lauryl alcohol) as fruit ripening agents can harm health. Studies indicate that chemicals alter the organic composition of vitamins and micronutrients while only changing the fruit's skin color, ensuring it remains raw inside [24, 25].

The date palm is a climacteric fruit that undergoes five distinct stages after pollination and fertilization: Hababouk (light green color), Kimri (immature green color), Khalal or Bisir (mature, red, or yellow color), Rutab (ripe, half flesh color brown), and Tamer (fully ripe, flesh completely brown color). Every stage is distinguished by variations in physical and biochemical characteristics [26]. Date palm trees produce fruits that do not ripen simultaneously, necessitating multiple pickings over several weeks. Harvesting ripe fruits at the right time is crucial for growers to maximize investment returns and obtain the highest quantity and quality of dates from each cultivar. The time taken to fruit ripening varies with each cultivar [27].

Moreover, unfavorable environmental conditions can cause some fruits not to ripen and remain red or yellow. These unripe fruits are vulnerable to insect pests and disease infestation, which cause further damage. When harvesting unripe fruits, most farmers either waste them or feed them to their animals. Therefore, artificial ripening methods are becoming simple, cost-effective, and environment friendly, reducing fruit production's ecological footprint. Unripe fruits of date palms are artificially ripened using ozone, table salt, vinegar, ultraviolet light, microwave, ultrasound, heat, and humidity [13, 28].

These artificial ripening techniques have some benefits, such as they emit less greenhouse gas emissions than traditional methods, thus reducing the environmental impact of the fruit industry. They can significantly reduce water consumption in the fruit industry by allowing fruits and vegetables to ripen without requiring water. Environmentally friendly ripening techniques can be utilized to ripen fruits without the need for plant growth regulators and chemical substances. These methods help to improve food safety and security. In addition, environmentally friendly artificial fruit ripening can improve quality, increase shelf life, and increase availability [29]. Artificial ripening ensures a consistent supply of fruits throughout the year, extending their shelf life and allowing farmers to distribute them to markets even during off-seasons, thus reducing dependence on specific seasons or geographical locations. It can reduce post-harvest losses by ensuring fruits reach their optimal ripeness before market transport, especially for perishable fruits with short shelf life. This helps farmers prevent premature spoilage and increase the chances of selling produce before wastage. Artificial ripening offers greater control over natural ripening processes, resulting in more uniform and consistent fruits, which is crucial for commercial purposes where consumers expect standardized products with predictable characteristics. Artificially ripened fruits enhance marketability by enhancing their attractive appearance and desirable qualities like color, texture, and flavor, as consumers are more likely to purchase visually appealing, uniformly ripe fruits. Artificially ripening fruits allow for earlier harvesting and transportation while maintaining their firmness, reducing damage risks, and allowing for longer shelf life. This allows farmers to reach distant markets and expand their customer base, as ripe fruits are more susceptible to bruising and spoilage. Artificial ripening techniques use controlled environments such as temperature and humidity, enabling better storage and distribution management. This helps farmers extend fruit shelf life and ensure optimal fruit delivery to consumers [23, 30].

4. Artificial ripening methods

Artificial ripening methods for dates can be categorized into chemical and physical methods. Among the chemical methods, commonly used substances include ethylene (C_2H_4) gas, which triggers ripening by stimulating ethylene production within the fruit; calcium carbide (CaC_2), releasing acetylene gas when in contact with moisture but raising safety concerns; ethanol (C_2H_6O) vapor initiates ripening, and ethephon ($C_2H_6ClO_3P$) that decomposes to release ethylene gas. Physical methods encompass hot air circulation, employing heat to expedite metabolic processes, steam treatment, steam for ripening and microbial control, radiation exposure through ionizing radiation, and solar energy in solar dryers or greenhouses to harness sunlight for faster ripening. **Figure 1** shows the chemical and physical methods of artificial ripening.

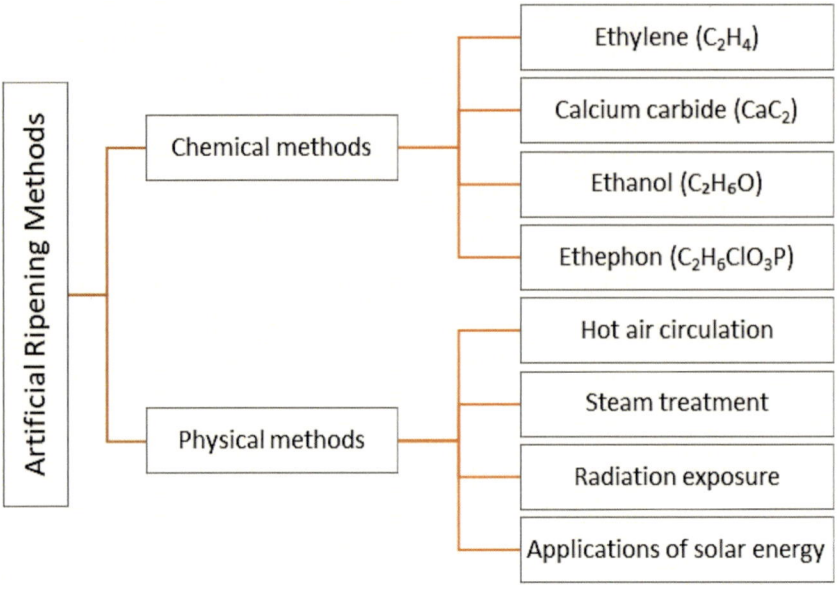

Figure 1.
Artificial ripening methods.

4.1 Chemical methods of artificial ripening

4.1.1 Artificial fruit ripening using ethylene (C_2H_4)

Climacteric and non-climacteric fruits are the main groups into which fruits are usually categorized. In general, climacteric fruits can ripen after harvest, while non-climacteric ones cannot. An elevated respiration rate and a subsequent burst of ethylene production are characteristics of climacteric fruit ripening. Fruit ripening is a complex, genetically programmed process that involves physiological, biochemical, and organoleptic changes, resulting in a soft, edible fruit with desirable quality attributes, possibly independent of each other. Ethylene, a naturally occurring hormone, is responsible for accelerating the ripening of fruits. It is synthesis from the activity of 1-aminocyclopropane-1-carboxylic acid (ACC) synthase (ACS) and 1-aminocyclopropane-1-carboxylic acid oxidase (ACO), which alter S-adenosyl-l-Met (SAM) into ACC and convert ACC into ethylene, respectively. The transmembrane receptor proteins found in the endoplasmic reticulum of cells sense ethylene [31, 32]. It can either inhibit or promote cell division, sometimes inhibiting cell expansion and sometimes stimulating lateral cell expansion. It triggers physiological, biochemical, and molecular changes in the fruit. It regulates fruit ripening by coordinating gene expression, causing respiration increase, autocatalytic ethylene production, and changes in color, texture, aroma, and flavor. The autocatalytic production of ethylene in climacteric fruits occurs when an initial concentration of ethylene increases, causing the fruit to naturally increase in signal production, accelerating the ripening process [33].

Globally, it is commercially produced through the steam cracking of petroleum hydrocarbons using various feedstocks such as ethane, propane, butanes, naphthas, and gas oils. Chemical treatment of ethylene for artificial fruit ripening involves exposing fruits to ethylene gas or using ethylene-releasing compounds such as ethephon, glyoxime, etacelasil, etc., used in the food industry to maintain a consistent supply of ripe fruits and meet consumer demands. One of the most popular techniques for artificial fruit ripening is ethylene gas treatment. Ethylene gas is introduced into a controlled atmosphere where fruits are stored. The type of fruit and the desired level of ripeness determine the ethylene gas concentration and exposure time. The gas penetrates through the fruit's skin and triggers a series of biochemical reactions that lead the fruit to ripen [34].

Ethylene-based artificial fruit ripening enhances fruit quality, appearance, and marketability by accelerating the ripening process and improving overall fruit characteristics. Artificial fruit ripening using ethylene ensures consistent ripeness across batches, giving consumers access to ripe fruits year-round. It also prevents variations in fruit taste, texture, and appearance. Exposing fruits to controlled amounts of ethylene gas can synchronize the ripening process, resulting in a more consistent product. This uniformity is crucial for commercial purposes, allowing better planning of harvesting, storage, transportation, and distribution [35]. Ethylene-induced artificial fruit ripening enhances flavor and aroma by promoting the synthesis of volatile compounds responsible for the characteristic flavors and aromas of ripe fruits. This process is particularly beneficial for fruits harvested prematurely or unripe for logistical reasons or to extend their shelf life [8, 36]. Ethylene-based artificial fruit ripening enhances visual appeal by stimulating color changes, resulting in vibrant hue angles and attractive pigmentation, which indicate fruit ripeness for consumers [37]. Artificial fruit ripening can reduce post-harvest losses by minimizing spoilage and extending fruit shelf life. Ethylene treatment accelerates ripening, allowing fruits to reach optimal consumption stages quickly. This is especially beneficial for short shelf life or perishable fruits. By reducing harvest-consumption time, ethylene treatment minimizes spoilage losses and increases fresh fruit availability in the market [35, 38]. Ethylene treatment has gained popularity because of consumer demand for fruits regardless of the time of year. Many consumers have developed tastes for fruits, which may not always be available in their area or at times of the year. These fruits can be harvested before they are completely ripe and artificially ripened with ethylene gas while being transported or stored. It enables vendors to satisfy customer requests and offer a steady supply of requested fruits all year round [39, 40].

It is important to highlight that while ethylene-induced artificial fruit ripening has many advantages, its application has potential drawbacks and concerns. Fruit quality and flavor can be negatively impacted by over-ripening or uneven ripening caused by incorrect application of ethylene or ethylene-releasing compounds. Fruits over-ripened, softened, or deteriorated due to improper application or overexposure to ethylene may have these effects. Food safety is of concern when artificial fruit ripening involves chemicals. Adherence to regulations and guidelines is imperative to ensure that the chemicals utilized are suitable for consumption and do not pose any health risks. To protect consumers and ensure food safety, ethylene use may also be subject to regulatory restrictions in certain countries or regions. Ethylene gas—a volatile organic compound (VOC)—can contribute to air pollution if not appropriately managed. Ethylene gas should be handled and disposed of safely to reduce any negative environmental effects [41, 42].

4.1.2 Artificial fruit ripening using calcium carbide (CaC_2)

Calcium carbide, a grayish-white crystalline solid produced by heating lime and coke in an electric furnace, is used in industries like steelmaking and welding due to its ability to generate acetylene gas (C_2H_2) when heated with water, a highly flammable fuel source. It is a chemical compound that reacts with water to produce acetylene gas and is used in artificial fruit ripening to accelerate the process; however, concerns have been raised about its safety and potential health risks. This controversial practice has been used in some regions globally. It is used for artificial fruit ripening to produce acetylene gas that triggers physiological changes such as fruit flesh softening, color development, and flavor enhancement. This mimics the effects of ethylene, a crucial plant hormone, allowing for the accelerated ripening process by exposing fruits to acetylene gas [43, 44]. Calcium carbide has potential risks, including extremely hazardous chemicals such as arsenic and phosphorus. These toxic chemicals contaminate fruits and pose serious health risks when consumed by humans, particularly arsenic, which can cause cancer and kidney diseases.

Similarly, the acetylene gas produced affects the neurological system by causing prolonged hypoxia [41]. The use of calcium carbide can cause irregular fruit ripening, causing unnatural appearance and imbalanced quality and potentially affecting the taste and nutritional content of fruits due to the accelerated ripening process. Therefore, due to these concerns, many countries have banned or strictly regulated the use of calcium carbide for artificial fruit ripening. Nonetheless, it is still in use in many countries, especially underdeveloped countries, where there is less awareness of the hazard risks. Alternative methods for artificial fruit ripening have been developed instead of relying on calcium carbide [45, 46].

4.1.3 Artificial fruit ripening using ethanol (C_2H_6O)

Ethanol, or ethyl alcohol, is a colorless, volatile liquid used in food and beverage production. Artificial fruit ripening is a practice that uses ethanol to expedite fruit ripening. Ethanol ripening is a promising technology for extending fruit shelf life, reducing food waste, improving flavor and quality, offering a sustainable alternative to traditional methods involving synthetic chemicals, and minimal equipment required. It is primarily used in fruit ripening as it stimulates ethylene production, a natural plant hormone responsible for regulating physiological processes in plants, including fruit ripening, senescence, and abscission [47, 48]. This method is beneficial for fruits harvested prematurely or require ripening. It allows growers to regulate fruit ripening timing to ensure optimal maturity before sale or consumption. It can potentially prevent cell wall degradation, resulting in firmer and more flavorful fruits. It decreases the respiration rate in fruits, potentially extending their shelf life and improving food security. Ethanol is commonly used for artificial fruit ripening in the commercial production of certain fruits. Unripen fruits are exposed to ethanol vapor or ethylene gas in specialized rooms, triggering a cascade of biochemical reactions that result in softening, color change, and flavor development. This process is commonly used when fruits are still green and firm for transportation [49, 50].

Ethanol can be used to ripen fruits in two ways: by exposing them to ethanol vapor in a sealed container, applying ethanol directly to the fruits through a solution, or spraying with an ethanol mist. This process allows the ethanol to evaporate and fill the container, allowing the fruits to ripen more quickly. Additionally, it is a simple and low-cost process, requiring only a sealed container or sprayer [51]. The ripening rate of fruits depends on the type of fruit and desired ethanol concentration. Higher concentrations and longer exposure times lead to faster ripening, but excessive ethanol can damage fruits and inhibit ripening [52]. Although the use of ethanol for artificial fruit ripening accelerates the process, it has disadvantages, such as residual ethanol on the surface of fruit can affect taste or pose health risks if consumed excessively. Therefore, producers must follow appropriate guidelines and regulations regarding ethanol use for fruit ripening [48, 53].

4.1.4 Artificial fruit ripening using ethephon ($C_2H_6ClO_3P$)

Ethephon, 2-chloroethylphosphonic acid, is the most popular synthetic plant growth regulator commonly used in the agricultural industry to accelerate fruit ripening, fruit coloring, fruit yield, germination, and flower induction by releasing ethylene gas when it comes into contact with moisture. Ethylene gas from ethephon stimulates fruit ripening by triggering physiological changes such as cell wall softening, starch breakdown, sugar production, and color development [54]. The method of applying ethephon for the artificial ripening of fruits is dipping or spraying the fruits in a diluted ethephon solution. Fruits are also artificially ripened in a sealed chamber with an ethephon generator. The type of fruit and its desired ripening stage determine the concentration and application method. Ethephon exposure is crucial for fruit ripening, with higher concentrations and longer exposure times resulting in faster ripening. However, excessive ethephon can damage fruits and inhibit ripening, so it is essential to balance these two factors [55, 56].

The ripening process starts when the fruit tissues absorb the ethephon solution and transform it into ethylene gas [57, 58]. Ethephon ensures uniform ripening of fruits, allowing consistent quality and appearance in commercial production. This is especially important for fruits that ripen unevenly. Fruits treated with ethephon have an extended shelf life due to delaying senescence and slowing decay, resulting in longer storage and transportation periods. Artificial fruit ripening using ethephon also meets market demands by supplying ripe fruits throughout the year, allowing better planning and management of production cycles [59, 60].

Using ethephon to ripen fruit artificially has raised concerns about potential health risks. Regulatory authorities set maximum residue limits for ethephon in various fruits, but proper application practices and adherence to these guidelines are crucial for consumer safety. Moreover, attention should be given to the optimal ethephon concentration and exposure time, which depend on the type of fruit and desired ripening rate. Ensuring uniform ripening is challenging due to uneven distribution of the ethephon solution or gas and uneven stacking of fruits. Safety concerns include the chemical's potential for skin and eye contact. Despite its safety, it is crucial to store and use ethephon in a well-ventilated area and avoid any body contact. Therefore, carefully considering and optimizing these parameters are essential for successful ethephon use [61, 62].

4.2 Physical methods of artificial ripening

A controlled application of heat and moisture for artificial fruit ripening is a widely used technique in the agricultural industry to ensure a consistent supply of ripe fruits throughout the year despite seasonal variations. In order to simulate the natural ripening process, artificial fruit ripening usually entails exposing the fruits to particular temperature and humidity levels. The main objective is to accelerate fruit ripening by increasing ethylene gas production, a naturally occurring plant hormone. Fruits undergo various physiological changes from ethylene, including softening, color development, and flavor enhancement [30, 63].

Artificial fruit ripening is regulated by food safety authorities in several countries to protect consumers. These regulations frequently outline acceptable ranges for temperature and humidity as well as the usage of ripening compounds that have been approved. Sometimes, humidity and heat are combined with ethylene gas to accelerate ripening. Its use is, nevertheless, closely controlled to avoid overexposure that can pose health hazards [64].

Heating is a crucial step in the artificial ripening of fruit because it speeds up ethylene production. Fruits are frequently exposed to warm temperatures for a specific duration, usually between 20°C and 50°C [13, 65]. The type of fruit and desired level of ripeness determine the precise temperature and duration; for example, a temperature of 50°C is recommended for artificial date palm fruit ripening [13]. Humidity control is also crucial for artificial fruit ripening, as it maintains firmness and prevents moisture loss. Low humidity can cause fruit desiccation and shriveling. The ideal humidity range for artificial fruit ripening varies depending on the fruit type and cultivar but typically ranges between 85–95% [13, 66].

Several techniques, such as hot air circulation, steam treatment, or radiation exposure, can achieve heating.

4.2.1 Hot air circulation

It is a widely used technique for artificial fruit ripening in agricultural and horticultural practices. It involves the controlled application of warm air to accelerate the process of fruits, maintaining consistent temperature and humidity levels. This method is primarily used in commercial settings such as fruit storage facilities, warehouses, and packing houses [67]. The process involves setting the desired temperature and humidity range, which vary with the type of fruit being ripened. Hot air circulation systems, consisting of fans or blowers, distribute warm air evenly throughout the ripening room, ensuring uniform conditions and equal exposure to heat, whereas humidity is provided using a humidifier. Regular monitoring and adjusting temperature and humidity levels is essential during the fruit ripening, allowing operators to adjust the hot air circulation system. Automated systems equipped with sensors and controllers can help streamline this monitoring process. This method ensures uniform ripening, accelerates the ripening process, and extends the shelf life of fruits [13, 68]. Mohammed and Alqahtani [30] designed a sensor-based artificial ripening system (S-BARS) combined with ultrasound pretreatment, an efficient approach for improving the quality of date fruits. **Figure 2** shows an image of the experimental setup of the sensor-based artificial ripening system. The system effectively controlled temperature and relative humidity, resulting in improved color and density of the artificially ripened fruits. The ultrasound pretreatment reduced the required time for ripening, decreased the percentage of damaged fruits, and increased the percentage of ripened fruits. The optimal treatment combination for ultrasound pretreatment and ripening parameters resulted in high-quality date fruits with attributes such as fruit weight, density, color, firmness, total soluble solids, pH, and sugars [30].

4.2.2 Steam treatment

The Steam treatment method involves exposing fruits to high temperatures and humidity to stimulate the ripening process. Steam treatment is a method used to accelerate fruit ripening by inactivating enzymes that inhibit it and activating those that promote it. It also increases the permeability of the fruit's cell walls, allowing ethylene gas to enter. A well-insulated steam chamber is used for this process, with the temperature and humidity set to optimal levels for the ripened fruit type. The steam

Figure 2.
Experimental setup of the sensor-based artificial ripening system.

circulates around the fruit, accelerating the ripening process. The ripening time varies depending on the fruit type but is typically shorter than at room temperature [69, 70].

4.2.3 Radiation exposure

Infrared radiation, a type of electromagnetic radiation, is widely used in various applications, including fruit ripening, accelerating the process, and enhancing the quality of fruits by using specific wavelengths. It also increases the permeability of the cell wall of the fruits, making it easier for ethylene gas to enter. It has longer wavelengths than visible light but shorter than radio waves. It is categorized into near-infrared (NIR), mid-infrared (MIR), and far-infrared (FIR) categories with distinct properties and applications. NIR radiation, with wavelengths ranging from 700 to 2500 nm, is primarily used for fruit ripening [71]. It penetrates the outer layers of fruits, absorbing energy and causing physiological changes that promote ripening in the fruit's tissues. The absorbed energy increases the fruit temperature, accelerating metabolic processes such as respiration and ethylene production. NIR also enhances enzymatic activity in fruits, causing color, flavor, aroma, and texture changes. It also influences the breakdown of complex carbohydrates into simpler sugars, making fruits sweeter [72].

Infrared radiation is a non-destructive and non-contact method for accelerating fruit ripening, unlike traditional methods that cause damage to the surface of fruit or structure. It offers precise control over the ripening process by adjusting the intensity and duration of radiation exposure. It benefits commercial fruit producers who must ensure consistent ripening across large quantities of fruits. It also enhances fruit quality by increasing antioxidant and vitamin levels and improving sensory attributes [73]. An IR ripening chamber must be well-insulated and equipped with Infrared emitters to use Infrared radiation for fruit ripening. The wavelength and intensity of the radiation should be set to the optimal ripening conditions for the fruit type. The Infrared radiation heats the fruit inside the chamber, accelerating the ripening process. The ripening time typically is shorter than the fruits ripened using hot air circulation or steam treatment methods [74]. Infrared radiation offers several advantages for fruit ripening, including faster ripening times. It helps to reduce transportation and storage costs, improve fruit quality by promoting uniform ripening, reduce food waste by increasing fruit shelf life, and increase profits for farmers and retailers. It also has a few disadvantages, including high installation and running costs, complex monitoring and control systems, and potential health risks to the eyes and skin. It is, therefore, essential to take appropriate safety precautions when using this technology, especially for more significant operations [75].

4.2.4 Application of solar energy for artificial fruit ripening

Solar energy is a renewable and sustainable energy source that can be used for heating, cooling, lighting, and powering appliances. It is also utilized for fruit ripening, a crucial process in the agricultural industry that impacts fruits' quality and shelf life. Solar energy can significantly reduce costs associated with fruit ripening by eliminating the need for fossil fuels. This clean and renewable energy source also minimizes the environmental impact of fruit ripening. Furthermore, solar energy can enhance the sustainability of the fruit industry by decreasing its reliance on nonrenewable fuel sources [76]. Solar energy can be utilized in various ways:

- The fruits can be dried using solar radiation to remove moisture, extending their shelf life and helping preserve them. This technique works especially well for drying heat-sensitive fruits. A solar dryer, a tray or basket with a clear plastic or glass cover over it to hold the fruits, is used for sun drying. The moisture content of fruit evaporates as heat from the sun's rays enters through the cover [77].

- In order to slow down the ripening process, solar storage involves storing fruits at a constant temperature using sun energy. This technique performs especially well for fruits that are sensitive to temperature changes. A solar-powered refrigeration system, which uses the sun's energy to cool the fruits, is also used for solar storage [78].

- Using sun energy to hasten fruit ripening is known as solar ripening. This technique works especially well for crops like bell peppers and tomatoes plucked before they fully mature. The process of solar ripening involves exposing the fruits to direct sunshine while they are covered with transparent plastic or glass. The fruits will mature more rapidly because the sun's rays penetrate the cover and heat them up. Unripe date palm fruits at the Khalal stage of development are harvested and artificially ripened using ambient solar energy [79].

- Solar-powered ventilation uses solar energy to maintain a controlled atmosphere for fruit ripening. Most fruits that require specific temperature and humidity levels benefit most from this strategy. To assist in keeping the fruits' environment stable, solar-powered ventilation also uses a dehumidifier or fan [80].

- The process of producing ethylene gas with solar energy is known as solar-powered ethylene production, which is used to ripen fruits. Sunlight is converted into ethylene gas using a solar-powered ethylene generator, which can be used to produce ethylene using solar power. Climacteric fruits that need a high ethylene concentration can benefit most from this strategy [81].

5. Effects of artificial ripening on data quality

Date palm fruit development undergoes a phase change due to ripening, which is influenced by primary and secondary metabolism changes. This process forms carotenoids, flavor compounds, antioxidants, and sugars, enhancing the nutritional quality of fruits [82]. Enzymes trigger these responses, causing changes in date palm fruit color, taste, and texture. Chlorophyll degradation leads to new pigments, while the amylase enzyme breaks down starch into sugar, making the fruit sweeter [83]. Due to climatic variability, date fruits in the Rutab stage may take a long time to reach the Tamar stage. Thus, in order to expedite the ripening process, date palm farmers commonly use artificial ripening techniques [57]. Artificial ripening can significantly impact the nutritional value of dates. According to certain studies, artificially ripened dates could have less antioxidants and vitamin C than naturally ripened dates. However, the nutritional value of naturally ripened and artificially ripened dates seems to be similar, according to other studies. Although the effects of artificial ripening using chemicals on the nutritional value of date palm fruit are not entirely known, some research has shown that it may be detrimental.

Artificial ripening by chemicals requires ethylene, calcium carbide, ethanol, ethephon, etc. These chemicals produce ethylene gas, which triggers fruit ripening in controlled environments. The dry weights of pulp and seed, titratable acidity, soluble solids, and respiration rates increased. In contrast, pH, firmness, and astringency decreased when the Shahani date palm cultivar fruits were treated with ethephon [84]. The application of ethephon reduced the fruit transition time from Kimri to Rutab. Also, it enhanced biochemical properties such as ascorbic acid, glucose, fructose, sucrose, total phenolics, total flavonoids, and total antioxidants of date palm cultivars Hillawi and Khadrawi [58]. El-Kafrawy and Abdel-Hamid [85] applied three techniques (sun drying, oven heat, and calcium carbide) for the artificial date palm cultivar Sewy ripening. They reported that although all three ripening methods improved fruit quality and reduced the decayed fruits and weight loss, sun drying artificial ripening was the best compared to others.

In addition to chemical artificial techniques, other non-chemical techniques that improved fruit texture, color, flavor, aroma, shelf life, etc., were also applied, including hot air circulation, steam treatment, and radiation exposure. In a study conducted on unripe date palm fruits of cultivar Dhakki, the application of sodium chloride proved to be more successful, leading to a 75% increase in ripening and enhanced fruit quality [86]. Similarly, when utilizing a 2% brine solution, Khalal fruits of the Dakkai cultivar can be artificially ripened to a level of up to 75% [87]. The artificial ripening of unripe Biser date palm fruits of cultivar Khalas was found to be significantly improved by a combination of temperature (50°C) and humidity (50%). As a result, weight loss and artificial ripening time were decreased while the fruit's marketable size, color, firmness, total soluble solids, pH, and sugars were improved [13]. Another study revealed that microwave pretreatment (80 W for 50 seconds) and controlled temperature (50°C) treatments significantly improve the nutritional profile and structural characteristics of date fruit cultivar Khupra, with microwave-processed samples being more acceptable than sun-dried ones [28]. The artificially ripened date fruits' color and density were improved by the ultrasound pretreatment, which also reduced the amount of time and electricity needed for fruit ripening and increased the percentage of ripened fruits without having an adverse effect on the fruit quality characteristics [30].

6. Regulatory and safety considerations

Food safety is a significant concern due to potential adverse impacts on human nutrition and health, including food-borne diseases, toxicants, zoonotic infections, agrochemical use, pesticide exposure, and antimicrobial resistance. Despite growing concerns in industrialized, high-income countries, developing countries experience the most damaging effects of unsafe foods, which bear the greatest burden of food-borne diseases. These diseases disproportionately affect malnourished individuals, children, pregnant women, and the elderly with weak immune systems, leading to a vicious cycle of morbidity and mortality [88]. Poor food safety issues hinder developing nations' agricultural development and access to export markets regulated by the World Trade Organization, including Sanitary and Phytosanitary Measures and Technical Barriers to Trade Agreements. Food safety significantly impacts developing countries' economies, although quantifying it in monetary terms is challenging [89].

Date palm growers use artificial ripening techniques in order to meet customer demand since natural ripening is a slow process. Artificial ripening, a method of accelerating the ripening of dates using chemicals and other physical methods, can enhance yields and reduce costs; however, it also raises concerns about food safety and consumer protection. Farmers and vendors generally use artificial fruit ripening agents, but their potential health hazards have led to global debate over the effectiveness of these methods. As such, strict regulations are in place to regulate the artificial ripening of dates to guarantee consumer safety and product quality. The regulatory requirements for the artificial ripening of dates vary from country to country [41]. Consuming acetylene directly can reduce brain oxygen supply and cause hypoxia. Calcium carbide, an alkaline compound, can cause stomach disorders after consuming artificially ripened fruits. Industrial-grade calcium carbide contains impurities like arsenic and phosphorus, leading to health risks like dizziness, frequent thirst, mouth and nose irritation, weakness, skin damage, difficulty swallowing, vomiting, and skin ulcers. Ethylene glycol consumption can cause kidney failure. These potential health risks highlight the need to carefully handle these products [24, 90, 91].

In most countries, the use of calcium carbide for artificial ripening is prohibited. The US Food and Drug Administration (FDA) and the Canadian Food Inspection Agency (CFIA) prohibit the use of calcium carbide for artificial ripening, deeming date palm fruits ripened with calcium carbide unsafe for human consumption and prohibiting their import or sale. In a few other countries, the use of calcium carbide is regulated but allows for artificial ripening. However, it is allowed only to be used by licensed facilities and following their guidelines. The artificially ripened fruits must be labeled for the consumers' information to make an informed choice [64, 92]. The FDA in the US regulates the artificial ripening of dates under the FD&C Act, ensuring they are safe and not adulterated, misbranded, or violating the Act. The agency has set guidelines for using artificial ripening agents in dates, including the types and maximum levels of use. California, one of the largest date-producing states in the United States, has regulations governing the use of artificial ripening agents in dates. The state prohibits the use of certain chemicals, such as ethylene gas, in the process, in addition to federal regulations [92].

Many countries have banned harmful chemicals like calcium carbide and potassium nitrate in the artificial ripening of dates due to potential health risks, including cancer and reproductive diseases. These chemicals have been linked to health problems, making their use in date production strictly regulated. The concerned food authorities mandate that all dates, whether naturally or artificially ripened, be labeled as "ripened" or "mature", indicating that the date has been treated with an artificial ripening agent, ensuring consumers understand the product's nature and potential benefits. Internationally, the guidelines for the artificial ripening of dates have been created by the Food and Agriculture Organization (FAO) and the World Health Organization (WHO) in collaboration with the Codex Alimentarius Commission, a joint food standard-setting authority. Countries may utilize the Codex principles as a foundation to create their own laws pertaining to the use of artificial ripening agents in dates [93].

The Gulf Co-operation Council (GCC) imports fresh fruits from other countries, comprising Bahrain, Kuwait, Oman, Qatar, Saudi Arabia, and the UAE. To export fruits to GCC countries, producers and exporters must comply with regulations set by these countries, ensuring fruit safety and quality. The Gulf Standards Organization (GSO), comprising six GCC countries and Yemen, aims to promote scientific and technical advancement in agricultural and food industries. Despite the development of around 1000 food-related legislations and standards, there are still differences between proposed standards and existing international guidelines. The GSO Food Standard Act mandates that fruits should be provided fresh to consumers upon preparation and packaging. The GSO Food Standards Committee harmonizes GCC standards with guidelines from Codex Alimentarius, ISO, and other international organizations, with most acts following the Codex Alimentarius. Gulf countries adhere to GSO standards collectively, but some countries have imposed additional acts under national acts for food safety [94].

7. Conclusion

Ripening is a process in fruits that enhances their edible qualities, resulting in a sweeter, less green, and softer fruit. The ripening process of many fruits, including date palm, can be very slow, uneven, and unpredictable, leading to the use of chemicals or physical methods to ripen fruits artificially. Ethylene and ethylene-generating chemicals like ethephon and calcium carbide accelerate ripening and improve peel color. However, these chemicals have some disadvantages in post-harvest shelf life and can harm product quality and human health. Ethylene is an explosive gas and expensive, whereas calcium carbide is banned in many countries, posing serious health hazards to humans. However, due to several health-related concerns, the effectiveness of chemically induced artificial ripening has come under criticism. The date palm fruits are also artificially ripened with a few non-hazardous chemicals, such as vinegar and salt, although this affects the flavor of the fruits. In order to ripe the unripe date fruits, alternative fruit ripening methods—which are easy to use, eco-friendly, and reasonably priced—have gained interest. Artificial heat and humidity treatments can be used to ripen date palm fruits when the tree is still not completed or when early rains threaten to damage the harvesting process. Among these physical artificial ripening methods are hot air circulation, steam treatment, and infrared radiation exposure.

Additionally, the oven-drying method can enhance the quality and sensory qualities of unripe date palm fruits. Similarly, in various horticulture-based enterprises, unripe date palm fruits are ripened in controlled conditions with different humidity and temperature ranges depending on the cultivar. Recently, date palm fruits subjected to microwave radiation shortened the ripening time of unripe fruits. It is expected that solar energy will be used to supply the energy needed to implement physical artificial ripening techniques. Because it is a renewable and sustainable energy source used for heating, cooling, lighting, and powering appliances, it will reduce costs, minimize environmental impact, and enhance the sustainability of the date palm fruit industry by decreasing its reliance on nonrenewable fuel sources. Moreover, throughout the artificial ripening process, real-time monitoring of the quality and sensory characteristics of date palm fruit will be performed via the use of artificial intelligence (AI) and Internet of Things (IoT) technologies. Additionally, these technologies will be used to automate a range of operations related to artificial ripening settings.

Acknowledgements

The authors gratefully acknowledge the financial support from Date Palm Research Center of Excellence, King Faisal University, Saudi Arabia.

Conflict of interest

The authors declare no conflict of interest.

Author details

Maged Mohammed[1,2*], Nashi K. Alqahtani[1,3] and Muhammad Munir[1]

1 Date Palm Research Center of Excellence, King Faisal University, Al-Ahsa, Saudi Arabia

2 Faculty of Agriculture, Agricultural and Biosystems Engineering Department, Menoufia University, Shebin El Koum, Egypt

3 Department of Food and Nutrition Sciences, College of Agricultural and Food Sciences, King Faisal University, Al-Ahsa, Saudi Arabia

*Address all correspondence to: memohammed@kfu.edu.sa

IntechOpen

© 2023 The Author(s). Licensee IntechOpen. This chapter is distributed under the terms of the Creative Commons Attribution License (http://creativecommons.org/licenses/by/3.0), which permits unrestricted use, distribution, and reproduction in any medium, provided the original work is properly cited.

References

[1] Chao CCT, Krueger RR. The date palm (Phoenix dactylifera L.): Overview of biology, uses, and cultivation. HortScience. 2007;**42**(5):1077-1082

[2] FAOSTAT. Food and Agriculture Organization of the United Nations. Database. Crop Production. Rome, Italy: FAOSTAT; 2021. Available from: https://www.fao.org/faostat/en/#data/QCL [Accessed: September 20, 2023]

[3] Aleid SM, Al-Khayri JM, Al-Bahrany AM. Date palm status and perspective in Saudi Arabia. In: Date Palm Genetic Resources and Utilization. Dordrecht: Springer Netherlands; 2015. pp. 49-95. Available from: https://link.springer.com/10.1007/978-94-017-9707-8_3

[4] Assirey EAR. Nutritional composition of fruit of 10 date palm (Phoenix dactylifera L.) cultivars grown in Saudi Arabia. Journal of Taibah University for Science. 2015;**9**(1):75-79

[5] Arias E, Hodder AJ, Oihabi A, Jimnez E, Hodder AJ, Oihabi A. FAO support to date palm development around the world: 70 years of activity. Emirates Journal of Food Agriculture. 2016;**28**(1):1-11. Available from: http://www.ejmanager.com/fulltextpdf.php?mno=204338

[6] Dhehibi B, Ben Salah M, Frija A. Date palm value chain analysis and marketing opportunities for the Gulf Cooperation Council (GCC) countries. In: Kulshreshtha SN, editor. Agricultural Economics - Current Issues. Rijeka, Kvarner, Croatia: IntechOpen; 2019. pp. 11-17. Available from: https://www.intechopen.com/books/agricultural-economics-current-issues/date-palm-value-chain-analysis-and-marketing-opportunities-for-the-gulf-cooperation-council-gcc-coun

[7] Statista. Leading Fruit Dates Exporters Worldwide in 2021. New York, USA: Statista Research Department, Statista Inc.; 2022. Available from: https://www.statista.com/statistics/961359/global-leading-exporters-of-dates [Accessed: September 20, 2023]

[8] El Hadrami A, Al-Khayri JM. Socioeconomic and traditional importance of date palm. Emirates Journal of Food and Agriculture. 2012;**24**(5):371-385

[9] Akram M, Kahlown MA, Soomro ZA. Desertification control for sustainable land use in the Cholistan Desert, Pakistan. In: The Future of Drylands. Dordrecht: Springer Netherlands; 2008. pp. 483-492. Available from: http://link.springer.com/10.1007/978-1-4020-6970-3_44

[10] Ahmed Mohammed ME, Refdan Alhajhoj M, Ali-Dinar HM, Munir M. Impact of a novel water-saving subsurface irrigation system on water productivity, photosynthetic characteristics, yield, and fruit quality of date palm under arid conditions. Agronomy. 2020;**10**(9):1265. Available from: https://www.mdpi.com/2073-4395/10/9/1265

[11] Aljaloud S, Colleran HL, Ibrahim SA. Nutritional value of date fruits and potential use in nutritional Bars for athletes. Food and Nutrition Sciences. 2020;**11**(6):463-480

[12] Mohammed M, Sallam A, Munir M, Ali-Dinar H. Effects of deficit irrigation scheduling on water use, gas exchange,

yield, and fruit quality of date palm. Agronomy. 2021;**11**(11):2256

[13] Mohammed M, Sallam A, Alqahtani N, Munir M. The combined effects of precision-controlled temperature and relative humidity on artificial ripening and quality of date fruit. Foods. 2021;**10**(11):2636. Available from: https://www.mdpi.com/2304-8158/10/11/2636

[14] Solomon S. Climate change 2007: The physical science basis. In: Contribution of Working Group I to the Fourth Assessment. Cambridge, United Kingdom/New York, NY, USA: Report of the Intergovernmental Panel on Climate Change. Cambridge University Press; 2007

[15] Kalra N, Kumar M. Simulating the impact of climate change and its variability on agriculture. In: Climate Change and Agriculture in India: Impact and Adaptation. Cham: Springer International Publishing; 2019. pp. 21-28. Available from: https://link.springer.com/10.1007/978-3-319-90086-5_3

[16] Piao S, Ciais P, Huang Y, Shen Z, Peng S, Li J, et al. The impacts of climate change on water resources and agriculture in China. Nature. 2010;**467**(7311):43-51

[17] Shabani F, Kumar L, Taylor S. Climate change impacts on the future distribution of date palms: A modeling exercise using CLIMEX. Magar V, editor. PLoS One. 2012;**7**(10):e48021. Available from: https://dx.plos.org/10.1371/journal.pone.0048021

[18] Zaid A, Wet D. Climatic Requirements of Date Palm. Rome, Italy: Food and Agriculture Organization Plant Production and Protection; 2002. Available from: http://www.fao.org/3/Y4360E/y4360e0c.html

[19] Faci M, Benziouche SE. Contribution to monitoring the influence of air temperature on some phenological stages of the date palm (cultivar 'Deglet nour') in Biskra. Journal of the Saudi Society of Agricultural Sciences. 2021;**20**(4):248-256

[20] Farooq S, Maqbool MM, Bashir MA, Ullah MI, Shah RU, Ali HM, et al. Production suitability of date palm under changing climate in a semi-arid region predicted by CLIMEX model. Journal of King Saud University – Science. 2021;**33**(3):101394. Available from: https://linkinghub.elsevier.com/retrieve/pii/S1018364721000550

[21] Mansoor M, Khalil SHK, Islam Z, Asif M, Akbar G, Khan MA, et al. Vulnerability of date palm cv. Dhakki to climate change and viable options for adaptation. Pakistan Journal of Agricultural Research. 2022;**35**(2):366-370

[22] Maheswaran S, Sathesh S, Priyadharshini P, Vivek B. Identification of artificially ripened fruits using smart phones. In: 2017 International Conference on Intelligent Computing and Control (I2C2). Coimbatore, India: IEEE; 2017. pp. 1-6. Available from: http://ieeexplore.ieee.org/document/8321857/

[23] Hewajulige IGN, Premaseela HDSR. Fruit ripening: Importance of artificial fruit ripening in commercial agriculture and safe use of the technology for consumer health. Sri Lanka Journal of Food and Agriculture. 2020;**6**(1):57-66

[24] Siddiqui MW, Dhua RS. Eating artificially ripened fruits is harmful. Current Science. 2010;**99**(12):1664-1668

[25] Mehnaz Mursalat, Asif Hasan Rony, Abul Hasnat, Mohammed Sazedur Rahman, Mohammed Nazibul Islam, Mohidus Samad Khan.

A Critical analysis of artificial fruit ripening: Scientific, legislative and socio-economic aspects. Chemical Engineering Science Management. 2013;**4**(1):1-7. Available from: www.chethoughts.com

[26] Alsmairat N, Othman Y, Ayad J, Al-Ajlouni M, Sawwan J, El-Assi N. Anatomical assessment of skin separation in date palm (Phoenix dactylifera L. var. Mejhoul) fruit during maturation and ripening stages. Agriculture. 2022;**13**(1):38. Available from: https://www.mdpi.com/2077-0472/13/1/38

[27] Mohamed RMA, Fageer ASM, Eltayeb MM, Mohamed Ahmed IA. Chemical composition, antioxidant capacity, and mineral extractability of Sudanese date palm (Phoenix dactylifera L.) fruits. Food Science & Nutrition. 2014;**2**(5):478-489

[28] Alvi T, Khan MKI, Maan AA, Shahid M, Sablani S. Microwaves as sustainable approach for artificial ripening of date fruit cv. Khupra to reduce fruit waste. Food Bioscience. 2023;**54**:102829. Available from: https://linkinghub.elsevier.com/retrieve/pii/S2212429223004807

[29] Hussain A, Pu H, Sun DW. Innovative nondestructive imaging techniques for ripening and maturity of fruits – A review of recent applications. Trends in Food Science and Technology. 2018;**72**:144-152. Available from: https://linkinghub.elsevier.com/retrieve/pii/S092422441730609X

[30] Mohammed M, Alqahtani NK. Design and validation of automated sensor-based artificial ripening system combined with ultrasound pretreatment for date fruits. Agronomy. 2022;**12**(11):2805. Available from: https://www.mdpi.com/2073-4395/12/11/2805

[31] Paul V, Pandey R, Srivastava GC. The fading distinctions between classical patterns of ripening in climacteric and non-climacteric fruit and the ubiquity of ethylene-an overview. Journal of Food Science and Technology. 2012;**49**(1):1-21

[32] Liu M, Pirrello J, Chervin C, Roustan JP, Bouzayen M. Ethylene control of fruit ripening: Revisiting the complex network of transcriptional regulation. Plant Physiology. 2015;**169**(4):2380-2390

[33] Abbas MF, Ibrahim MA. The role of ethylene in the regulation of fruit ripening in the Hillawi date palm (Phoenix dactyliferaL). Journal of the Science of Food and Agriculture. 1996;**72**(3):306-308

[34] Barry CS, Giovannoni JJ. Ethylene and fruit ripening. Journal of Plant Growth Regulation. 2007;**26**(2):143-159

[35] Lobo MG, Yahia EM, Kader AA. Biology and post-harvest physiology of date fruit. In: Dates: Post-Harvest Science, Processing Technology and Health Benefits. Sussex, UK: Wiley; 2013. pp. 57-80. Available from: https://onlinelibrary.wiley.com/doi/10.1002/9781118292419.ch3

[36] Lim S, Lee JG, Lee EJ. Comparison of fruit quality and GC–MS-based metabolite profiling of kiwifruit 'Jecy green': Natural and exogenous ethylene-induced ripening. Food Chemistry. 2017;**234**:81-92. Available from: https://linkinghub.elsevier.com/retrieve/pii/S0308814617307422

[37] Younuskunju S, Mohamoud YA, Mathew LS, Mayer KFX, Suhre K, Malek JA. Genome-wide analysis of dry (Tamar) date palm fruit color. bioRxiv. 2022;**12**:1-26. Available from: http://biorxiv.org/content/early/2022/12/14/2022.12.12.520041.abstract

[38] Bleecker AB, Kende H. Ethylene: A gaseous signal molecule in plant. Annual Review of Cell and Developmental Biology. 2000;**16**(1):1-18

[39] Chervin C, El-Kereamy A, Roustan JP, Latché A, Lamon J, Bouzayen M. Ethylene seems required for the berry development and ripening in grape, a non-climacteric fruit. Plant Science. 2004;**167**(6):1301-1305

[40] Pech JC, Purgatto E, Bouzayen M, Latché A. Ethylene and fruit ripening. Plant Hormone Ethylene. 2012;**44**:275-304

[41] Ur-Rahman A, Chowdhury FR, Alam MB. Artificial ripening: What we are eating. Journal of Medicine. 2008;**9**(1):42-44

[42] Janssen S, Schmitt K, Blanke M, Bauersfeld ML, Wöllenstein J, Lang W. Ethylene detection in fruit supply chains. Philosophical Transactions of the Royal Society of London. Series A, Mathematical and Physical Engineering Sciences. 2014;**372**(2017):20130311. Available from: https://royalsocietypublishing.org/doi/10.1098/rsta.2013.0311

[43] Nura A, Dandago A. Atu N', Wali R. Effects of artificial ripening of banana (Musa spp) using calcium carbide on acceptability and nutritional quality. Journal of Post-Harvest Technology. 2018;**6**(2):14-20. Available from: http://www.jpht.info

[44] Gomes FR, DFP DS, Costa GS, PHM DS, Da Silveira-Neto AN, SCS C. Calcium carbide in anticipation and standardization of ripening in Cajá-manga fruits. Revista Brasileira de Fruticultura. 2023;**45**:139. Available from: http://www.scielo.br/scielo.php?script=sci_arttext&pid=S0100-29452023000100302&tlng=en

[45] Okeke ES, Okagu IU, Okoye CO, TPC E. The use of calcium carbide in food and fruit ripening: Potential mechanisms of toxicity to humans and future prospects. Toxicology. 2022;**468**:153112. Available from: https://linkinghub.elsevier.com/retrieve/pii/S0300483X22000245

[46] Oladipupo A, Coker HAB. The use of calcium carbide in fruit ripening: Health risks and arsenic index as a quantitative marker for calcium carbide residue potentials excipients from natural sources view project Chinwendum ALARIBE view project. Progress in Chemical and Biochemical Research. 2022;**5**(2):125-132. Available from: www.pcbiochemres.com

[47] Nichols WC, Patterson ME. Ethanol accumulation and Poststorage quality of 'delicious' apples during short-term, low-O_2, CA storage. HortScience. 2022;**22**(1):89-92

[48] Dudley R. Ethanol, fruit ripening, and the historical origins of human alcoholism in primate Frugivory. Integrative and Comparative Biology. 2004;**44**(4):315-323. Available from: https://academic.oup.com/icb/article-lookup/doi/10.1093/icb/44.4.315

[49] Pesis E. The role of the anaerobic metabolites, acetaldehyde and ethanol, in fruit ripening, enhancement of fruit quality and fruit deterioration. Postharvest Biology and Technology. 2005;**37**(1):1-19

[50] Umesh AR, Venkatesh HN, Kiragandur M, Mohana DC. Artificial ripening of fruits—Misleading ripe and health risk. Everyman's Science. 2016;**6**(1):364-369

[51] Jowkar MM, Rahmanian AR, Zakerin A. Artificial ripening of 'Shiraz'persimmon (Diospyros kaki Thunb. cv.'shiraz') prior to marketing.

International Journal of Fruit Science. 2007;**6**(4):13-24

[52] Karabulut OA, Gabler FM, Mansour M, Smilanick JL. Post-harvest ethanol and hot water treatments of table grapes to control gray mold. Postharvest Biology and Technology. 2004;**34**(2):169-177

[53] Conde C, Silva P, Fontes N, Dias ACP, Tavares RM, Sousa MJ, et al. Biochemical changes throughout grape berry development and fruit and wine quality. Food, Global Sciences Books Lyd, UK; 2007;**1**(1):1-22. Available from: http://hdl.handle.net/1822/6820

[54] Bhadoria P, Nagar M, Bharihoke V, Bhadoria A. Ethephon, an organophosphorous, a fruit and vegetable Ripener: Has potential hepatotoxic effects? Journal of Family Medicine and Primary Care. 2018;**7**(1):179

[55] Ban T, Kugishima M, Ogata T, Shiozaki S, Horiuchi S, Ueda H. Effect of ethephon (2-chloroethylphosphonic acid) on the fruit ripening characters of rabbiteye blueberry. Scientia Horticulturae (Amsterdam). 2007;**112**(3):278-281

[56] Mahajan BVC, Tajender K, Gill MIS, Dhaliwal HS, Ghuman BS, Chahil BS. Studies on optimization of ripening techniques for banana. Journal of Food Science and Technology. 2010;**47**(3):315-319

[57] Awad MA. Increasing the rate of ripening of date palm fruit (Phoenix dactylifera L.) cv. Helali by preharvest and post-harvest treatments. Postharvest Biology and Technology. 2007;**43**(1):121-127. Available from: https://linkinghub.elsevier.com/retrieve/pii/S0925521406002134

[58] Hussain I, Ahmad S, Amjad M, Ahmed R. Ethephon application at kimri stage accelerates the fruit maturation period and improves phytonutrients status (Hillawi and Khadrawi (C.V.)) of date palm fruit. Pakistan Journal of Agricultural Sciences. 2015;**52**(2):415-423

[59] Meena R, Yadav PK, Singh RS. Responseof date palm cultivars to varying concentration of Ethephon on yield and yield attributing characters. Progressive Agriculture. 2013;**13**(1):1-3

[60] Dhillon WS, Mahajan BVC. Ethylene and ethephon induced fruit ripening in pear. Journal of Stored Products and Postharvest Research. 2011;**2**(3):45-51. Available from: https://academicjournals.org/journal/JSPPR/article-full-text-pdf/604E9DD8533

[61] Alvarez F, Arena M, Auteri D, Binaglia M, Castoldi AF, Chiusolo A, et al. Peer review of the pesticide risk assessment of the active substance ethephon. EFSA Journal. 2023;**21**(1):7742. Available from: http://doi.wiley.com/10.2903/j.efsa.2023.7742

[62] Royer A, Laporte F, Bouchonnet S, Communal PY. Determination of ethephon residues in water by gas chromatography with cubic mass spectrometry after ion-exchange purification and derivatisation with N-(tert-butyldimethylsilyl)-N-methyltrifluoroacetamide. Journal of Chromatography. A. 2006;**1108**(1):129-135

[63] Mahmood MH, Sultan M, Miyazaki T. Significance of temperature and humidity control for agricultural products storage: Overview of conventional and advanced options. International Journal of Food Engineering. 2019;**15**(10):20190063.

Available from: https://www.degruyter.com/document/doi/10.1515/ijfe-2019-0063/html

[64] Islam MN, Mursalat M, Khan MS. A review on the legislative aspect of artificial fruit ripening. Agriculture & Food Security. 2016;**5**(1):8. Available from: http://agricultureandfoodsecurity.biomedcentral.com/articles/10.1186/s40066-016-0057-5

[65] Cho BH, Koseki S. Determination of banana quality indices during the ripening process at different temperatures using smartphone images and an artificial neural network. Scientia Horticulturae (Amsterdam). 2021;**288**:110382. Available from: https://linkinghub.elsevier.com/retrieve/pii/S0304423821004891

[66] Weichmann J. Post Harvest Physiology of Vegetables. Vol. 145. New York, NY, USA: Marcel Bekker Inc.; 1987. 145 p

[67] Alsmairat N, Al-Qudah T, El-Assi N, Mehyar G, Gammoh I, Othman YA, et al. Effect of drying process on physical and chemical properties of "medjool" date palm fruits. Fresenius Environmental Bulletin. 2019;**28**(2A):1563-1570

[68] Lurie S. Post-harvest heat treatments of horticultural crops. Horticultural Reviews (Am Soc Hortic Sci). 2010;**22**(22):91-121

[69] Fallik E. Prestorage hot water treatments (immersion, rinsing and brushing). Postharvest Biology and Technology. 2004;**32**(2):125-134

[70] Hazbavi I, MHH K, Mostaan A, Banakar A. Effect of post-harvest hot-water and heat treatment on quality of date palm (cv. Stamaran). Journal of the Saudi Society of Agricultural. 2015;**14**(2):153-159. Available from: https://linkinghub.elsevier.com/retrieve/pii/S1658077X13000453

[71] Torricelli A, Contini D, Mora AD, Martinenghi E, Tamborini D, Villa F, et al. Recent advances in time-resolved NIR spectroscopy for nondestructive assessment of fruit quality. Chemical Engineering Transactions. 2015;**44**:43-48

[72] Marques EJN, De Freitas ST, Pimentel MF, Pasquini C. Rapid and non-destructive determination of quality parameters in the "Tommy Atkins" mango using a novel handheld near infrared spectrometer. Food Chemistry. 2016;**197**:1207-1214

[73] Rahmawati L, Saputra D, Sahim K, Priyanto G, Pan Z. Study of using infrared radiation for increasing the shelf life of duku. In: IV Asia Symposium on Quality Management in Post-Harvest Systems. Jeonju, Korea: ISHS Acta Horticulturae; 2017. pp. 109-116

[74] Kathirvelan J, Vijayaraghavan R. An infrared based sensor system for the detection of ethylene for the discrimination of fruit ripening. Infrared Physics & Technology. 2017;**85**:403-409. Available from: https://linkinghub.elsevier.com/retrieve/pii/S1350449517302190

[75] Loaharanu P, Ahmed M. Advantages and disadvantages of the use of irradiation for food preservation. Journal of Agricultural and Environmental Ethics. 1991;**4**(1):14-30

[76] Mekhilef S, Faramarzi SZ, Saidur R, Salam Z. The application of solar technologies for sustainable development of agricultural sector. Renewable and Sustainable Energy Reviews. 2013;**18**:583-594. Available from: https://linkinghub.elsevier.com/retrieve/pii/S1364032112006156

[77] Bennamoun L. Reviewing the experience of solar drying in Algeria with presentation of the different design aspects of solar dryers. Renewable and Sustainable Energy Reviews. 2011;**15**(7):3371-3379

[78] Ahmadi A, Ehyaei MA, Doustgani A, El Haj AM, Hmida A, Jamali DH, et al. Recent residential applications of low-temperature solar collector. The Journal of Cleaner Production. 2021;**279**:123549. Available from: https://linkinghub.elsevier.com/retrieve/pii/S0959652620335940

[79] Yahia EM, Lobo MG, Kader AA. Harvesting and post-harvest Technology of Dates. In: Dates: Post-Harvest Science, Processing Technology and Health Benefits. Chichester, UK: John Wiley & Sons Ltd; 2013. pp. 105-135. Available from: https://onlinelibrary.wiley.com/doi/10.1002/9781118292419.ch5

[80] Patel DP, Jain K, Lakhawat S, Wadhawan N, Jain SK, Scholar P, et al. Conceptual study of solar-powered evaporative cooling Systems for the Storage of different perishables-a significant appraisal. Environment and Ecology. 2023;**41**(2A):931-938

[81] Lanzafame P, Abate S, Ampelli C, Genovese C, Passalacqua R, Centi G, et al. Beyond solar fuels: Renewable energy-driven chemistry. ChemSusChem. 2017;**10**(22):4409-4419

[82] Awad MA, Al-Qurashi AD, Mohamed SA. Biochemical changes in fruit of an early and a late date palm cultivar during development and ripening. International Journal of Fruit Science. 2011;**11**(2):167-183

[83] Rastegar S, Rahemi M, Baghizadeh A, Gholami M. Enzyme activity and biochemical changes of three date palm cultivars with different softening pattern during ripening. Food Chemistry. 2012;**134**(3):1279-1286

[84] Rouhani I, Bassiri A. Effect of Ethephon on ripening and physiology of date fruits at different stages of maturity. Journal of Horticultural Sciences. 1977;**52**(2):289-297

[85] El-Kafrawy TM, Abdel-Hamid N. Effect of some artificial ripening treatments used as environmentally safe on fruit quality and storage ability of "Sewy" dates. In: Arab Palm Conference, Proceedings of the First International Scientific Conference for the Development of Date Palm and Dates Sector in the Arab World, 4-7 December 2011. Riyadh, Kingdom of Saudi Arabia. King Abdulaziz City for Science and Technology (KACST); 2011. pp. 147-162

[86] Saleem SA, Baloch AK, Baloch MK, Baloch WA, Ghaffoor A. Accelerated ripening of Dhakki dates by artificial means: Ripening by acetic acid and sodium chloride. Journal of Food Engineering. 2005;**70**(1):61-66. Available from: https://linkinghub.elsevier.com/retrieve/pii/S0260877404004352

[87] Saleem SA, Saddozai AA, Asif M, Baloch AK. Impact of artificial ripening to improve quality and yield for the export of "DHAKKI" dates. Acta Horticulturae. 2010;**882**:1125-1134. Available from: https://www.actahort.org/books/882/882_130.html

[88] Fung F, Wang HS, Menon S. Food safety in the 21st century. Biomedical Journal. 2018;**41**(2):88-95. Available from: http://link.springer.com/10.1007/BF02727155

[89] World Health Organization [WHO]. Estimates of the global burden of foodborne diseases. In: Food Borne Disease Burden Epidemiology Reference Group 2007-2015. Geneva, Switzerland: WHO; 2015. Available from: https://www.who.int/multi-media/details/estimates-of-the-global-burden-of-

foodborne-diseases%0Ahttps://www.who.int/multi-media/details/estimates-of-the-global-burden-of-foodborne-diseases-african-region

[90] Goonatilake R. Effects of diluted ethylene glycol as a fruit-ripening agent. Global Journal of Biochemistry and Biotechnology. 2008;**3**(1):8-13

[91] Fattah SA, Ali MY. Carbide ripened fruits—A recent health hazard. Faridpur Medical College Journal. 2010;**5**(2):37

[92] Capogrossi KL, Calvin M, Coglaiti D, Hinman S, Karns A, Lasher T, et al. Food Processing Sector. NC, USA: RTI International; 2015. pp. 1-24

[93] Food and Agriculture Organization of the United. Codex Alimentarius. FAO. Available from: https://www.fao.org/fao-who-codexalimentarius/news-and-events/en/?cat=74278 [Accessed: September 20, 2023]

[94] Taha M. United Arab Emirates - Food and Agricultural Import Regulations and Standards. Dubai, UAE: Office of Agricultural Affairs (OAA); 2013

Chapter 7

Impact of PEF (Pulsed Electric Fields) on Olive Oil Yield and Quality

Oleksii Parniakov, Sam David Hopper and Stefan Toepfl

Abstract

Olive oil holds significant importance in the European diet and is renowned globally for its sensory attributes and health benefits. The effectiveness of producing olive oil is greatly influenced by factors like the maturity and type of olives used, as well as the milling techniques employed. Generally, mechanical methods can extract approximately 80% of the oil contained in the olives. The rest 20% of the oil remains in the olive waste generated at the end of the process. Additionally, significant amounts of bioactive compounds like polyphenols are also lost in the olive pomace. Traditionally, heat treatment, enzymes, and other chemicals are used for the enhancement of oil extraction; however, this approach may impact the quality of olive oil. Therefore, new technology, such as pulsed electric field (PEF), is of great benefit for nonthermal yield and quality improvements.

Keywords: electroporation, olives, EVOO, extraction, PEF, bioactives

1. Introduction

The International Olive Council reports that olive oil production on a global scale has tripled over the past 60 years. For the 2021/22 crop year, it is estimated that production reached 3,098,500 tons [1].

Virgin olive oil (VOO) and especially extra virgin olive oil (EVOO) is considered as a main part of the Mediterranean diet. Furthermore, thanks to its rich content of bioactive compounds like phenols, biophenols, phytosterols, and tocopherols, regular consumption of olive oil enhances antioxidant levels and improves blood lipid profiles, thus lowering the risk of certain degenerative diseases like atherosclerosis and cancer [2]. The extraction of VOO and EVOO consists only of mechanical processes. According to the EU Regulation EC 1513/2001 [3], olive oil should consist only of the following steps: (i) the crushing of olives to break the plant tissues, facilitating the release of oil, (ii) the process of malaxation applied to the olive paste to encourage the coalescence of oil droplets, the temperature of olive paste at this stage should not exceed 27°C and usually done for around 1 h, and (iii) the mechanical recovery of the oil by decantation/centrifugation or pressing. Afterward, obtained oil/water mixture goes through speed separation and lately filtered or decanted to remove any possible, solid residues prior to bottling.

The efficiency of traditional EVOO and VOO production is relatively low, resulting in around 20% of oil losses. There are two main sources of such high losses: (i) oil remains inside the olive cells or (ii) it is emulsified with water. Those losses could be connected to the way and conditions of extraction or/and olives variety. Traditionally, thermal treatments or the addition of technological coadjuvants (talc, calcium carbonate, enzymes) have been used to improve the oil extraction from olives [4]. It is well known that a temperature rise reduces the viscosity of oil and improves its diffusivity, thus enhancing oil yield. However, olive oil obtained at temperatures higher than 27°C cannot be named EVOO. High temperatures also destroy thermosensitive compounds, such as phenols and aromatic compounds, that directly impact the organoleptic properties of oil. The usage of technological coadjuvants, in general, improves the oil yield and oil quality. For example, the usage on enzymes such as cellulases, hemicellulases, and pectinases can increase the oil yield by 1–2%. The level of the effect depends on the cultivar and olive fruit maturity [5]. Moreover, their use can increase the concentration of phenolic compounds in the paste and oil [6]. However, the purchase price of such adjuvants is quite high, in the range of 30–50 €/L.

Nowadays, emerging nonthermal technologies such as pulsed electric fields (PEF) and ultrasound are gaining more and more interest in food science and technology.

Pulsed electric fields (PEF) consist of the application of high voltage (1–30 kV) and short pulses (ns-μs) to the food products. The PEF is based on the mechanism of electroporation that, due to the electrocompression forces, induces the pores of different shapes and diameters on the cell membrane, hence improving mass and heat transfer processes. Therefore, the combination of conventional malaxation with PEF generally improves extraction efficiency and release of bioactive compounds. Moreover, its application can lead to improved nutritional and sensory quality of EVOO.

The increase in olive oil yield due to PEF (pulsed electric field) can be attributed to two factors: (i) the enhancement of oil extraction from olive tissue and (ii) the liberation of olive oil that was previously trapped in oil/water emulsion [7]. The first one could be explained by the electroporation of the cytoplasmic membranes that results in the improvement of mass transfer processes through cell envelopes [8, 9]. Therefore, PEF assists in the release of oil from lipo-vacuoles of mesocarp cells that have not been disrupted by crushing [10]. However, some part of oil is emulsified with water, and subsequently, it is lost with olive pomace after decantation process [11]. The main difficulty of freeing this bound oil is connected to the fact that droplets of emulsified oil are surrounded by a lipoprotein membrane [12]. Nevertheless, some authors hypothesized that the PEF treatment could have disrupted the lipoprotein membranes, thereby facilitating the release of oil [8].

Different research groups have investigated the influence of PEF on olive oil yield in the lab, pilot, and industrial scale [8, 13–15]. They have found that the olive oil yield can be enhanced by up to 15% with application of PEF. Moreover, the increase of the concentration of bioactive compounds such as polyphenols, phytosterols, and total tocopherols by 11.5, 9.9, and 15%, respectively, in oil extracted from PEF-treated olives can be achieved [8].

This chapter will discuss the effect of PEF treatment on plant tissue. Furthermore, a deep dive into the benefits of PEF application in the production of olive oil will be discussed. The process and quality benefits will be presented.

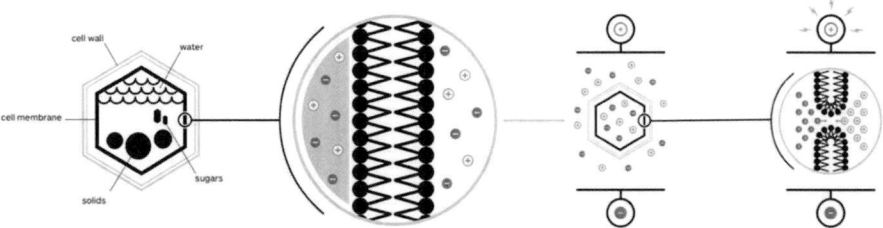

Figure 1.
Schematic representation of cell electroporation.

2. Action of PEF

PEF is based on mechanism of electroporation, and this phenomenon has been intensively researched for several decades in the scope of microbial inactivation and mass transfer enhancement [16, 17].

The applied electric field causes a dielectric breakdown of the membrane and thus influences the membrane structure and increases its permeability [18]. The electroporation process can be divided into four main steps [19]. Due to its nature biological cell membrane is considered as a capacitor filled with dielectric material of low electrical conductivity [17]. A cell membrane consists of phospholipid bilayer; therefore, it has a natural transmembrane potential, referring to the voltage difference by intra- and extracellular ionic concentrations [20]. Firstly, the electric field applied to the food product causes polarization of the cell membrane, leading to an increase in the transmembrane potential [21]. Secondly, hydrophilic pores are formed in the membrane. This pore formation stage can be explained by the electromechanical model, which is the most common one [21]. When the applied electric field surpasses a specific critical threshold, electrocompression forces emerge on both sides of the membrane, leading to the deformation and instability of the cell membrane (**Figure 1**).

Thirdly, the number and size of the pores increases with the duration of the electric pulse [21]. Depending on the intensity of the applied electric field, thickness of the membrane, cell shape, and size, electroporation can be reversible or irreversible. Usually, exceeding the critical electric field strength between 1 and 2 kV/cm induces irreversible pores in plant cells [22].

3. EVOO oil yield increase

Typical layout of the olive oil mill is presented in **Figure 2**. Usually, production of EVOO consists of crushing, malaxing, solid/liquid, and liquid/liquid separations.

PEF system is usually installed after the crasher to enhance the subsequent release of oil during malaxation based on the electroporation of the cell membranes. However, it can be implemented in some cases after the malaxation step to facilitate the demulsification of the oil and to improve mass transfer during the subsequent mechanical separation. Application of PEF in olive oil production has been intensively researched by different scientific groups at lab, pilot, and industrial scales [8, 13, 14]. Example of PEF industrial installation after crusher is presented in **Figure 3**. Here, 30 kW PEF system from Elea Technology GmbH has been used for treatment of 12 t/h of olive paste.

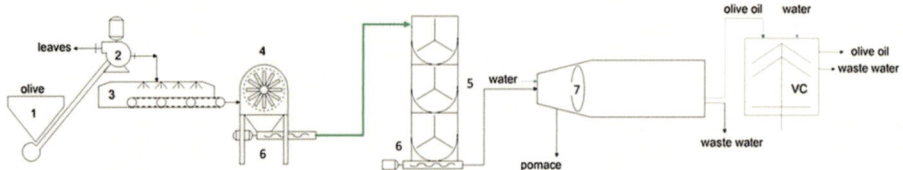

Figure 2.
Typical layout of the olive oil mill. Here, 1 is a loading hopper; 2 is a defoliator; 3 is a washing machine; 4 is a crusher machine; 5 is a malaxer; 6 is cavity pump stators; 7 is a solid/liquid horizontal centrifugal decanter. Adopted from [13].

Figure 3.
Industrial PEF installation. Own unpublished picture.

Research conducted at laboratory and industrial scales has demonstrated that the PEF treatment of the olives results in enhancement of oil extraction without negative impact on sensorial attributes while maintaining the EU legal standards for the highest quality EVOO [7].

The first studies on application of PEF on lab scale have indicated that application of this technology can increase the yield of olive oil by up to 7.4% (**Figure 4**) [15].

Research study conducted by [8] has mentioned that PEF treatment of olive fruits (Arroniz variety) resulted in increase of extraction yield by 2.7% mass compare to traditional method of extraction. Similar results have been obtained by application of PEF treatments to the olive paste from different olives varieties (Nocellara del Belice). It has been stated that positive effect on extractability was achieved after application of PEF treatment at 2 kV/cm and 7.83 kJ/kg. In fact, the yield has been increased by 6% when a PEF system was used. The authors stated that the PEF treatment of olive paste resulted in a significant reduction of oil loss in pomace of about 40.5% [23]. Furthermore, three different varieties of olive fruits (Tsounati, Amfissis, and Manaki variety) were subjected to PEF conditions ranging from 1.6 to 70.0 kJ/kg before being subjected to malaxation for 30 minutes at 30°C. As a result of this PEF treatment, the extraction yield of olive paste was boosted by up to 18 times compared

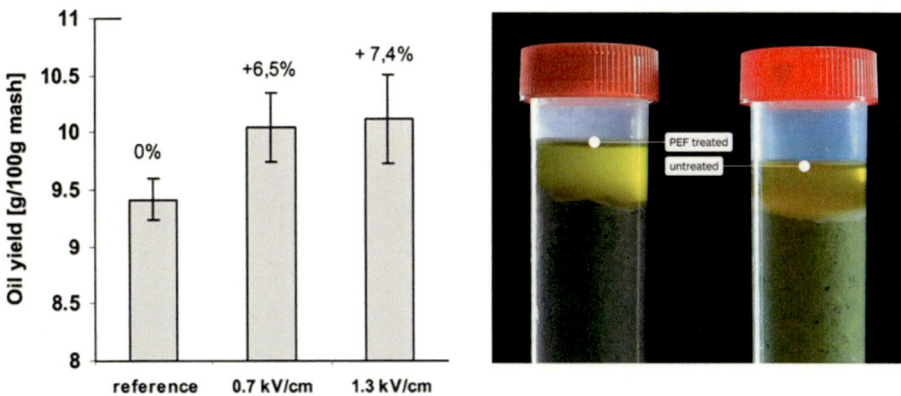

Figure 4.
Yield of oil from PEF-treated and untreated olive paste. Adopted from [15] and visual representation of oil yield increase.

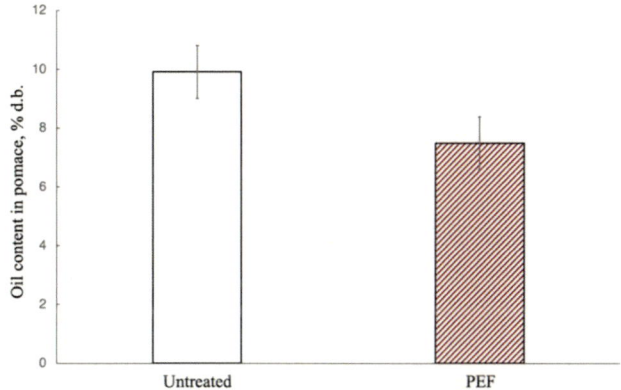

Figure 5.
Oil content in the untreated and PEF-treated olive pomace after solid/liquid separation. Adopted from [13].

to the untreated samples [24]. A recent study has shown that the pomace samples treated with PEF have a significantly lower content of residual olive oil than untreated samples (**Figure 5**). The authors stated that the extraction of olive oil has been significantly enhanced by more than 3% [13].

Additionally, to the enhancement of EVOO extraction, the reduction of malaxation time and temperature has been observed. It has been demonstrated that PEF implantation and malaxation at a low temperature of 15°C for only 30 min can increase oil yield by 3%, compared to the traditional thermal process at 26°C. It is worth mentioning that extraction at such low temperature reduces losses of volatile and bioactive compounds, which are important for the sensory quality of olive oil [14]. However, a more recent study, where the extraction efficiency has been evaluated for untreated and PEF-treated olive paste at different malaxation time, stated that there is a certain optimum time when maximal oil yield can be achieved. The authors found that for PEF-treated samples, 30 min of malaxation time was enough to reach maxima yield (**Figure 6**). At this time, an extra yield of 1.9 ± 0.3% was obtained for PEF samples compared to the untreated one. However, when a longer time of malaxation was used, these yield differences were reduced [25].

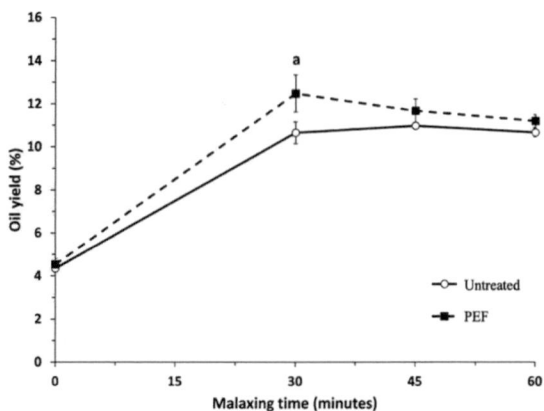

Figure 6.
Yield of oil from untreated and PEF-treated olive paste malaxed at different temperature [25].

4. Quality of PEF-extracted olive oil

EVOO is characterized by the high content of various bioactive compounds such as biophenols, tocopherols, and phytosterols [26]. The principal group of antioxidants in EVOO is hydrophilic phenols. These compounds are extremely important cause they determine the sensory characteristics, like bitterness, pungency, and stability of the oil [27], as well as determining the aroma and flavor of EVOO. The majority of bioactive compounds are thermosensitive, and their concentration in the final oil depends on many factors such as variety of fruits, harvesting time, process conditions, etc., [28]. Additionally, to increase oil yield PEF also improves the extraction of bioactive compounds from the olive paste. One research group has demonstrated that the concentration of bioactive compounds such as polyphenols, phytosterols, and total tocopherols in oil extracted from PEF-treated olives was higher compared to traditionally extracted one by 11.5, 9.9, and 15%, respectively [8]. In the traditional extraction process, the total tocopherol concentration in the oil was measured at 19.2 mg/100 g. However, with the application of pulsed electric field (PEF), there was a significant increase in tocopherol content by 15.0%, reaching a value of 22.0 mg/100 g. Notably, when comparing individual isomers in both traditional and PEF-extracted oils, only a-tocopherol showed a significant difference, with its value increasing by 25%, from 11.4 mg/100 g to 14.3 mg/100 g, respectively [8]. Another researcher found a slight increase of a-tocopherol as well in oil extracted from PEF-treated Arbequina olive pomace (1.67%) [14]. It was found an increase of total phenolic compounds (TPC), oleacein, and 3,4-DHPEA-EA in the oil obtained from the PEF-treated olive paste in the last phase of malaxation could probably be linked to enhanced PC release and solubilization [29]. Other researchers have also reported higher levels of phenolic compounds and oleuropein derivatives, particularly 3,4-DHPEA-EDA, in comparison to the control extra virgin olive oil (EVOO). However, there were no significant differences observed in the content of other phenolic compounds like secoiridoid derivatives (especially ligstroside), lignans, and α-tocopherol [13]. It has been noted that the enhanced extraction capability of this process justifies the energy requirements, making Pulsed Electric Field (PEF) technology a promising method for extracting oil from the olive paste matrix, with positive effects on the quality of the final products. Moreover, a slightly higher content of dialdehydic forms of decarboxymethyl elenolic acid linked to hydroxytyrosol (3,4-DHPEA-EDA or oleacein) and to tyrosol (p-HPEA-EDA or oleocanthal) was observed in EVOO extracted from Nocellara del Belice variety of olives using PEF technology compared to the control test [23].

Organoleptic properties of olive oil have high importance for the final consumers. Several studies have shown that PEF EVOO has a slightly higher bitterness compared to traditionally extracted. These findings can be explained by a generally higher content of polyphenols in the PEF EVOO. However, two other parameters such as fruitiness and spiciness are not affected by PEF [13].

One of the research groups was investigating the shelf-life of PEF and control EVOO. The tests suggest that the quality of oil derived from nonthermally pretreated olives is influenced by the chosen technology and process conditions. Notably, the application of pulsed electric field (PEF) treatment has shown to result in an increase in the oxidative stability of the oil obtained from these olives [24].

5. Conclusions

PEF treatment is one of the most promising emerging technologies. Numerous research groups have been investigating the effect of PEF on production of olive oil. It has been found that the application of PEF to the olive pomace prior to malaxation leads to various process and product benefits. It was demonstrated that oil could be easier extracted from PEF-treated olive paste. In general, the oil yield increase has been found in the range of 3–15%. Such a high range of yield increase might be connected to different olives varieties, harvesting conditions, and time, as well as processing parameters, mainly malaxation time and temperature. What is more, PEF can reduce the malaxation time by up to 30 min. This process benefit would help oil producers to increase their production capacities with relatively low investments and small footprints. It is well known that PEF increases the diffusion coefficient; moreover, the lower the process temperature is, the bigger this increase. Meaning, in the possibility of reducing the malaxation temperature that would lead to lower energy consumption and higher quality of the final product. Finally, PEF increases the extraction of health promoting compounds such as phenols, tocopherols, etc., resulting in healthier and more premium EVOO. Overall, it has been demonstrated that PEF treatment can be a perfect substitution for any type of used adjuvant (talc, enzymes, etc.). In the long run, implementation of PEF technology in the olive mill would remove the running costs connected to the costs of technological adjuvants and would open the possibility of increasing production capacities without a substantial capital investment.

Apart from technological advantages, implementation of PEF brings along extra monetary benefits. According to the aforementioned results the implementation of PEF treatment has the potential to raise the oil yield from 20% to 22.7%. In an industrial oil mill processing 160 tons of olives per day, the application of PEF treatment could increase the daily production of extra virgin olive oil (EVOO) by 4.33 tons, resulting in a totals output of 36.33 tons per day, up from the original 32 tons. According to the International Olive Council, the EVOO price in 2022 were between 2 and 3.7 €/kg [30], which would result in a potential extra turnover of between 8660 € and 16,021 € per production day. Additionally, the possible rise in the concentration of bioactive compounds, such as polyphenols or phytosterols, could positively influence the final price of the oil.

Conflict of interest

The authors declare no conflict of interest.

Author details

Oleksii Parniakov, Sam David Hopper and Stefan Toepfl*
Elea Technology GmbH, Quakenbrueck, Germany

*Address all correspondence to: s.toepfl@elea-technology.com

IntechOpen

© 2023 The Author(s). Licensee IntechOpen. This chapter is distributed under the terms of the Creative Commons Attribution License (http://creativecommons.org/licenses/by/3.0), which permits unrestricted use, distribution, and reproduction in any medium, provided the original work is properly cited.

References

[1] Nardella M, Moscetti R, Bedini G, Bandiera A, Chakravartula SSN, Massantini R. Impact of traditional and innovative malaxation techniques and technologies on nutritional and sensory quality of virgin olive oil – A review. Food Chemistry Advances. 2023;**2**:100163

[2] Cicerale S, Conlan XA, Sinclair AJ, Keast RSJ. Chemistry and health of olive oil phenolics. Critical Reviews in Food Science and Nutrition. 2009;**49**:218-236

[3] The European Parliament. Regulation (EC) No 1513/2001. 2001

[4] Squeo G, Difonzo G, Summo C, Crecchio C, Caponio F. Study of the influence of technological coadjuvants on enzyme activities and phenolic and volatile compounds in virgin olive oil by a response surface methodology approach. LWT. 2020;**133**:109887

[5] Polari JJ, Wang SC. Comparative effect of hammer mill screen size and Cell Wall-degrading enzymes during olive oil extraction. ACS Omega. 2020;**5**(11):6074-6081

[6] Vierhuis E, Servili M, Baldioli M, Schols HA, Voragen AGJ, Montedoro. Effect of enzyme treatment during mechanical extraction of olive oil on phenolic compounds and polysaccharides. Journal of Agricultural and Food Chemistry. 2001;**49**(3):1218-1223

[7] Cristina S-GA, Álvarez I. Applying pulsed electric fields to improve olive oil extraction. In: Javier R, Heinz V, editors. Pulsed Electric Fields Technology for the Food Industry: Fundamentals and Applications. Cham: Springer International Publishing; 2022. pp. 357-368. DOI: 10.1007/978-3-030-70586-2_11

[8] Puértolas E, Martínez De Marañón I. Olive oil pilot-production assisted by pulsed electric field: Impact on extraction yield, chemical parameters and sensory properties. Food Chemistry. 2015;**167**:497-502

[9] Carbonell-Capella JM, Parniakov O, Barba FJ, Grimi N, Bals O, Lebovka N, et al. "Ice" juice from apples obtained by pressing at subzero temperatures of apples pretreated by pulsed electric fields. Innovative Food Science and Emerging Technologies. 2016;**33**:187-194. Available from: https://www.scopus.com/inward/record.uri?eid=2-s2.0-84954348848&partnerID=40&md5=e52161c679ea3af1adf3d4a82813b9c8

[10] Chiacchierini E, Mele G, Restuccia D, Vinci G. Impact evaluation of innovative and sustainable extraction technologies on olive oil quality. Trends Food Science and Technology. 2007;**18**(6):299-305. Available from: https://www.sciencedirect.com/science/article/pii/S092422440700043X

[11] Aguilera MP, Beltran G, Sanchez-Villasclaras S, Uceda M, Jimenez A. Kneading olive paste from unripe 'Picual' fruits: I. Effect on oil process yield. Journal of Food Engineering. 2010;**97**(4):533-538. Available from: https://www.sciencedirect.com/science/article/pii/S0260877409005809

[12] Espínola F, Moya M, Fernández DG, Castro E. Improved extraction of virgin olive oil using calcium carbonate as coadjuvant extractant. Journal of Food Engineering. 2009;**92**(1):112-118. Available from: https://www.sciencedirect.com/science/article/pii/S0260877408005311

[13] Leone A, Tamborrino A, Esposto S, Berardi A, Servili M. Investigation on the effects of a pulsed electric field (PEF) continuous system implemented in an industrial olive oil plant. Food. 2022;**11**(18):2578

[14] Abenoza M, Benito M, Saldaña G, Álvarez I, Raso J, Sánchez-Gimeno AC. Effects of pulsed electric field on yield extraction and quality of olive oil. Food and Bioprocess Technology. 2013;**6**:1367-1373

[15] Guderjan M, Töpfl S, Angersbach A, Knorr D. Impact of pulsed electric field treatment on the recovery and quality of plant oils. Journal of Food Engineering. 2005;**67**(3):281-287

[16] Barba F, Parniakov O, Wiktor A. Pulsed Electric Fields to Obtain Healthier and Sustainable Food for Tomorrow. London, United Kingdom: Academic Press; 2020. p. 352

[17] Zimmermann U, Pilwat G, Riemann F. Dielectric breakdown of cell membranes. Biophysical Journal. 1974;**14**(11):881-899

[18] Toepfl S, Siemer C, Heinz V. Effect of high-intensity electric field pulses on solid foods. In: Emerging Technologies for Food Processing. San Diego: Academic Press; 2014. pp. 147-154

[19] Saulis G, Venslauskas MS. Cell electroporation. Bioelectrochemistry and Bioenergetics. 1993;**32**(3):221-235

[20] Yeagle PL. Chapter 5 - Membrane Models. In: Yeagle PL, editor. The Membranes of Cells. Third Edittion ed. Boston: Academic Press; 2016. pp. 85-94

[21] Saulis G. Electroporation of cell membranes: The fundamental effects of pulsed electric fields in food processing. Food Engineering Reviews. 2010;**2**(2):52-73

[22] Toepfl S, Heinz V, Knorr D. Overview of pulsed electric field processing for food. In: Emerging Technologies for Food Processing. London: Academic Press; 2005. pp. 69-97

[23] Tamborrino A, Urbani S, Servili M, Romaniello R, Perone C, Leone A. Pulsed electric fields for the treatment of olive pastes in the oil extraction process. Applied Sciences. 2020;**10**(1):114

[24] Andreou V, Dimopoulos G, Alexandrakis Z, Katsaros G, Oikonomou D, Toepfl S, et al. Shelf-life evaluation of virgin olive oil extracted from olives subjected to nonthermal pretreatments for yield increase. Innovative Food Science & Emerging Technologies. 2017;**40**:52-57

[25] Martínez-Beamonte R, Ripalda M, Herrero-Continente T, Barranquero C, Dávalos A, de las Hazas MC, et al. Pulsed electric field increases the extraction yield of extra virgin olive oil without loss of its biological properties in Frontiers in Nutrition 2022;2022, 9

[26] Jimenez-Lopez C, Carpena M, Lourenço-Lopes C, Gallardo-Gomez M, Lorenzo JM, Barba FJ, et al. Bioactive compounds and quality of extra virgin olive oil. Food. 2020;**9**(8)

[27] Sánchez de Medina V, El RM, Priego-Capote F, Luque de Castro MD. Mass spectrometry to evaluate the effect of the ripening process on phenols of virgin olive oils. European Journal of Lipid Science and Technology. 2013;**115**(9):1053-1061

[28] Gimeno E, Castellote AI, Lamuela-Raventós RM, De la Torre MC, López-Sabater MC. The effects of harvest and extraction methods on the antioxidant content (phenolics, α-tocopherol, and β-carotene) in virgin olive oil. Food Chemistry. 2002;**78**(2):207-211

[29] Veneziani G, Esposto S, Taticchi A, Selvaggini R, Sordini B, Lorefice A, et al. Extra-Virgin olive oil extracted using pulsed electric field technology: Cultivar impact on oil yield and quality. Frontiers in Nutrition. 2019;**6**

[30] International Olive Council. Olive oil prices – June 2022 update. 2022. pp. 1-30. Available from: https://www.internationaloliveoil.org/wp-content/uploads/2022/06/IOC-prices-rev-0-1.html#fn1